Being and becoming

Jose Luis Perez Velazquez • Vera Nenadovic

Being and becoming

A guide to act in the theatre of existence

Springer

Jose Luis Perez Velazquez
The Ronin Institute
Montclair, NJ, USA

Vera Nenadovic
Lawrence Bloomberg School of Nursing
University of Toronto
Toronto, Canada

ISBN 978-3-030-78263-4 ISBN 978-3-030-78264-1 (eBook)
https://doi.org/10.1007/978-3-030-78264-1

This Springer imprint is published by the registered company Springer Nature Switzerland AG
The registered company address is: Gewerbestrasse 11, 6330 Cham, Switzerland

Preface

As I have witnessed throughout my years people striving against the frustrations and other predicaments of life, I have often wondered whether science could help relieve the state of mind that accompanies those dissatisfactions. Science has indeed provided enormous benefit to society, from medicine to technology. We live in a world to a large extent shaped by science. But the alleviation of the distressed state of mind is more personal than that which can be provided with technology, formally or professionally. Naturally, one can always visit a psychiatrist or psychologist to deal with the life's burdens, but the majority of us do not need such professional assistance. During my career as neuroscientist and biophysicist, I have gathered an integrated view on natural phenomena which has helped me comprehend what life may be all about. This has helped me obtain a relief, liberation if you will, from the chains of our created life. I wrote this book to share with others what science is teaching us about some specific aspects in this matter. Much of the content is based in the graduate course "Integrative perspectives on consciousness and self-awareness" taught at the University of Toronto. While this is a scientific text, it is aimed at a lay audience. It is written with minimal technical jargon and references to books and scientific papers enough to satisfy the curious who want more detail. In current times, with a vast flood of scientific papers and data, it is convenient, even totally necessary, to find some order on an apparently disjoint cohort of papers from a variety of sources, trying to extract the key points. This "simplification" is one aim of this book, and my colleague and co-author Vera Nenadovic has contributed to make some specific notions more understandable to lay audiences and as well to "translate" some of the scientific findings and their "take home messages" to our current everyday living. Some parts contain particular details about experiments, but they are provided for completeness, to acknowledge the science and explore some aspects that are difficult to translate into lay terms. These more technical parts are not needed to understand the main theme. And because some of the described scientific observations share some common ground with the teachings of philosophical and spiritual traditions, there is a part on philosophy and religion as these relate to the specific topics presented in the text.

Oviedo, Asturias, 2021

Montclair, NJ, USA Jose Luis Perez Velazquez
Toronto, Canada Vera Nenadovic

Contents

Introductory remarks, or why you should read this book

"El triunfo supremo de la razón [...] es poner en duda su propia validez" (The supreme triumph of reason [...] is to cast doubt on its own validity)
Miguel de Unamuno, *Del sentimiento trágico de la vida* (1954)

"Nothing in life is to be feared, it is only to be understood. Now is the time to understand more, so that we may fear less"
Maria Skłodowska-Curie

It seems to me most people spend considerable time seeking a sense of purpose in life and, concomitant with that, a sense of personal identity. Perhaps this is an attempt to alleviate insecurities or maybe to elevate themselves to higher grounds. In our opinion this search is a fallacy. In this book we will describe why we believe such a search is a delusion. To start, we only offer a short answer: the personal identity, the so-called self, is a perception, an illusion created by the ultimate illusionist—your embodied brain. And the purpose in life, for organisms like us who can enjoy experiences, is not much more than to precisely enjoy the experience.

The intention of this book is to inform, perhaps enlighten, readers about results in several sciences that may guide us to some sort of liberation. A relief that is needed in the current time and age when mankind has managed to create a tremendously complex life with a diverse burden of chores. A deliverance from some of the most poignant afflictions that prevent individuals from achieving the true purpose of our lives: the enjoyment of the experience in the present, not in the past or the future. Among these afflictions we find two primordial concerns: the belief and subsequent attachment to a self, and the conviction that life must have a deep purpose in which we are major players.

The scientific observations we will review, along with considerations from philosophical perspectives and religious traditions, will perchance reveal a path to liberation from the chains of the "self", that construction which makes us so vulnerable to insults, criticisms or unmet expectations, liberation from the anguish of the potential insignificance of our lives and from the distress caused by the delusions of disappointments.

The roads to liberation will consist of observations in the natural sciences, more particularly in neuroscience but also in physics. We believe it is time that the impact of science is felt not only in the technological side – the popular perspective of the benefit of science – but also in the informational aspect, the teachings that science can offer us to improve our lives through a deeper understanding of those two major searches mentioned in this book's introductory sentence. We will also take advantage of philosophical views and the teachings of religious traditions that will support the scientific perspectives and guide us towards that desired liberation.

While this book is aimed at a lay audience, we will have to indulge in some science. Don't be intimidated, we will translate. And for readers interested in more technical details, there are specific sections and many references that can be inspected as desired. These sections or the scientific papers – there is a vast body of literature on the topics treated here and only a tiny, infinitesimal portion is presented in the reference list at the end of the text – will not be really necessary to understand the matters treated. Thus, the organization of the book aims, in its first part on neuroscience, at improving the understanding of that personal identity, that self that all people in all cultures construct – and as we shall see, other animals too. Because as it will become apparent, the self is a construction made up by the brain. We will see how the brain is the ultimate illusionist. We will see that almost everything that we term reality is made up in the space between our ears and above our eyes. The journey through neuroscientific observations will reveal how resilient this illusion of a personal identity is. In spite of it being about scientific findings, hopefully it will not be a boring trip: it will not consist of a detailed description of – to many readers – boring scientific experiments and analyses, rather the essence of the scientific results will be distilled. In fact, the trip will be quite entertaining. We shall see the immense power of brains to alter our bodies specially in pathologies like dissociative identity disorder – previously known as multiple personality disorder. We will witness how the two halves of our brain may have an independent personality, or how one can have a near-death experience – like an out-of-body experience – without actually being near death, but rather sitting comfortably in a laboratory of cognitive science. We will also meet self-aware worms and slime mould mental prodigies.

The second part of the book shows the perspective from physics, especially biophysics. It will illustrate that, as much as the appearance of the self is an inevitable occurrence in organisms like us with a complex nervous system, life too is an inevitable phenomenon on a planet with the characteristics of Earth. As in the first part, scientific observations will be summarised, while keeping them accessible; you will embark on a fascinating journey. You will learn how to perform a miracle, that is, what seems a very improbable event, and for which you need only a dual compartment box separated and some little beads. Intrigued? Discovering the science of the apparent miracle helps us comprehend the nature of phenomena and why things happen; why the ants in your garden follow that trail and not another, or why lightning may strike you but not your neighbour. You may even find out why there is something instead of nothing; this may or may not help in the final goal of mitigating life's afflictions but you can take it as a bonus question-answer for reading this book. The third part of the book will reveal how some philosophical and religious traditions share common themes with the notions derived from the scientific approaches.

The overarching theme is encapsulated in the title of the book. It is the deep understanding that everything changes: being and becoming. "From Being to Becoming" is the title of a book by Ilya Prigogine, Nobel prize in chemistry in 1977 (From Being to Becoming: Time and Complexity in the Physical Sciences, W. H. Freeman & Co Ltd., 1980). His work on non-equilibrium thermodynamics paved the route for the understanding of how patterns emerge in nature and how order arises from disorder. These notions are not only pertinent to the understanding of chemistry but to many other aspects such as the description of incidents in social systems. Like the rest of natural phenomena, we are now something, and we become something else later. Everything changes, especially in the brain, as exposed in Marvin Minsky's words, "But it makes no sense to speak of brains as though they manufacture thoughts the way factories make cars. The difference is that brains use processes that change themselves and this means we cannot separate such processes from the products they produce [...] The principal activities of brains are making changes in themselves" (Minsky, 1985). A true, in-depth understanding of this seemingly trivial fact starts the aforementioned road towards liberation. Accepting the inevitability of change, which entails accepting the inevitability of death, is a beginning of our deliverance from the grief and frustration when we witness the transience of things. Let us start exploring, in Part I, how that persistent perception termed the self is and becomes, and what this has to say about our happy or unhappy lives.

Part I
The Neuroscience Perspective – Where Is My Self?

The nature of reality is experienced as determined by the information-processing capabilities of our embodied nervous systems – that is, nervous systems integrated in a body. For animals like us, the brain is the key component, one of the two parts of the central nervous system (the other being the spinal cord). But we have to understand, as this book will show readers, that the brain in isolation without the rest of the nervous system, the body and the environment around us, would not do much. Hence when the word "brain" is used throughout this book, it is in reality a shorthand for "the embodied nervous system immersed in an environment". The basic neuroscience that Part I will cover will make this statement clear. We are all familiar with the experience of embodiment, and statements like "he has a good heart", which comes from ancestral times, is an illustration of interconnectedness between brain and other organs, such that when the brain is aware of, say, dangerous or distressing things, it influences heart activity via the vagus nerve and the heart beating starts to speed up. It is via the nerves and biochemical compounds like hormones that the heart is coupled to the nervous system. Hence this sense of embodiment prompted our ancestors to endow the heart with cognitive powers. Now that we know more about human physiology, perhaps we should change that sentence above to "he has a good brain", although admittedly it sounds less poetic than the original.

While the processing of sensory inputs at the level of the sensory organs (e.g. retina in the eyes or the organ of Corti of the cochlea in the ears) operates near-optimal performance and the response properties of sensory neurons seem to be optimal for the encoding of stimuli, the processing at later stages within the brain may lose that optimality, so what we finally perceive may not be what is in front of us. Why would optimally performing sensory organs and a normal brain not lead to optimal perception? Limitations exist at several levels, imposed by neurophysiology as well as by psychology, from the unreliability of cell-to-cell contact through synapses (as explained in Sect. 1.4, synapses are the contact between neurons) to the restrictions of the mental functions derived from the psychological level. It is a fundamental goal of neuroscience to understand the limitations of information processing in nervous systems, and particularly in the mammalian brain, not only to better comprehend cognitive functions and the perception of reality but also to characterise pathological brain states. Equally important is another of the main goals of contemporary neuroscience, the understanding of "self-consciousness", that is, that the brain is not only aware of its surroundings but also of its own functioning. This scheme contains the implicit assumption that brain states like the "self" can be comprehended by the brain "itself". A multitude of results in the neurosciences demonstrate how constraints imposed by the autonomous neural activity and internal brain architecture and those imposed by the environment, in addition to the limits of the brain's logical capabilities, brings about the experience of reality. Among these experiences, one stands out in all individuals: self-awareness, or the perception of the self, the personal identity.

Chapter 1
The Emergence of the Self

Contents

Our society is a cultural medium for the development of selves and/or egos. It will incessantly continue to breed and feed each individual's self. Now that the word ego has appeared, let us clarify one point. While the words self and ego may be considered synonymous, the term self will be used in this book due to two main reasons. First, the self is a clearer neuroscientific term investigated in experimental neuroscience, while ego has other implications in psychology, psychiatry and related fields. Second, the term ego has a very specific, mostly negative, connotation for most people: saying someone has an ego is many times interpreted as this person being selfish, egocentric, egomaniac, etc. But having a self, which as we will see is an inevitable consequence for animals like us, does not always imply selfishness or egocentrism; hence, the preferred term that will be used is self and its plural, selves.

From an early age, in their small worlds of home and caregivers, children are encouraged to take pride and see themselves as worthy individuals. Later on in life, in the larger world, some will receive prizes, words of praise and other things that will contribute to an enlargement of their egos (yes, here, we can use the word ego if only to facilitate readers' understanding of what is meant) with the usual counterparts like excessive pride. Similarly, in the larger world of social media, the self

J. L. Perez Velazquez, V. Nenadovic, *Being and becoming*,
https://doi.org/10.1007/978-3-030-78264-1_1

compares to carefully constructed and curated selves many times leading to disappointment or purchasing of paraphernalia thought to achieve a better sense of self. The nurture of a sense of the personal identity is not that bad, provided there was a deep understanding of what that self is or is not. The problem is that without this knowledge, it may get out of hand. In some individuals with propensity to egocentrism, being exposed to the aforementioned continuous praise and similar actions taken in our societies may lead to trouble, developing an inflated ego that in the end will make them very vulnerable to criticisms or insults. In others, not necessarily egocentric, there could be an experience of frustration, sorrow and bitter disappointment when the achievements are not those expected. Related to this enhanced vulnerability of large selves, probably most readers will have noticed that in this time and age, it is not difficult to do or say something that will be perceived as insulting by some. Why are people so susceptible? Prominent selves are the major reason for this extreme susceptibility. In fact, we would go so far as to say that it is the only reason, because if there is no self—that is, there is nobody in your head—who can be insulted or offended? As a Buddhist mentor once said, "No self, no problem". And along this line of reasoning, a minimal sense of self can make us invulnerable to failure, because there would not be anybody in our head to feel disappointed, and of course, this applies too to the other side of the coin—success! Rudyard Kipling in his poem "If" echoed this idea: "If you can meet with triumph and disaster and treat those two impostors just the same".[1]

In general, a large self makes the individual vulnerable; vulnerable to disapproval and criticism by others, to frustration due to events gone the wrong way against expectations. It is of interest that the personality traits, the features associated with the self, may be different depending on the culture in which one develops. This may be true in the case of being part of an individualist or a collectivist culture; in the former, personal achievement and uniqueness of the individual are prominent characteristics that are reinforced, whereas in the latter, the group guiding behaviours seem more important. Thus, these distinct aspects may promote the emergence of selves with different traits; nonetheless, these are still selves. There are further issues with developing selves, and this is why comprehending the nature and rise of selves is fundamental; for instance, it is known that emergent selves (that is, personalities) of children raised in abusive environments do not expect anything but negative experience—punishment, not reward. But of course, one thing is the perception of a self, of a unity in cognition and behaviour, and another thing is the several personality traits that accompany this perception. In general, once the seeds of the self are planted, its growth is inevitable, even without help from parents or educators. Animals too have a perception of their own identity, as anyone who has had pets can attest, although they may not have had educators fostering their achievements. But before we proceed to understand why the emergence of the self is an unavoidable consequence of an embodied nervous system in a changing environment, by going back to the basics and the beginning—i.e. primitive animals—, let us start to cast some doubt on the feeling of our selves.

[1] www.poetryfoundation.org

1.1 Perception Is Everything

Let us advance a possible definition of the self that will be substantiated in the following sections: the self can be considered as a creation based on the perception of unity and continuity in behaviour and cognition. If it is a perception, a consideration of what these perceptions are, how they form, in other words, what it is that we perceive, is in order. The physicist David Bohm offers us a clue about what may be going on when we perceive that which we term reality, when he said, "Reality is what we take to be true. What we take to be true is what we believe. What we believe is based upon our perceptions. What we perceive depends on what we look for. What we look for depends on what we think. What we think depends on what we perceive. What we perceive determines what we believe. What we believe determines what we take to be true. What we take to be true is our reality".

I challenge you to try to catch yourself in an instance with no perceptions. As the philosopher David Hume said in his 1740 book *Treatise of Human Nature*, "I can never catch myself at any time without a perception, and never can observe anything but the perception" (Hume, 1740). Same with me, and I am sure you will agree too. You may close your eyes, ears, nose and any other (sensory) orifice you wish, but you will still be perceiving something. Even during sleep, we perceive things. The most obvious case is during rapid eye movement (REM) sleep (the period of sleep when normally dreams take place) when there are dreams, moments when you will be perceiving an internal world created by your brain. You can see, hear, smell, taste and feel in dreams without any external input to your sensory organs. And if you happen to have a lucid dream, it is even more manifest that you are perceiving things/events because lucid dreams are dreams in which the subject is aware that it is a dream (Voss et al., 2009). As an aside, there are relatively easy methods to try to induce lucid dreams which are really fun because you can do anything you wish, like flying, and yet being completely aware that it is not real, rather a dream—I used the (possibly the simplest) method to evoke lucid dreams, thinking just before going to sleep that I will have a lucid dream, that I will become aware that I am dreaming. On some occasions, I was able to have it and used it to do what I enjoy most during dreams: I was flying at will, which was extremely amusing to be able to "consciously" fly how and where I wanted. This is particularly fascinating in that in our "real" world, we have never experienced flight (that is, without technology), and yet in the lucid dream, we understand that we are flying and its sensation.

Going back to perceptions, on the whole, it can be said that perceptions constitute the mental world; nothing exists but observations and sense impressions. It can be distinguished whether the perceptions become conscious or not, and we should be careful asserting that a stimulus that we do not perceive consciously has not been perceived at all by our brain. Indeed, subliminal messages widely used in media like TV or in stores prove that some parts of our brains do perceive stimuli that do not enter our awareness, but yet these stimuli can change our behaviour. In other words, each region of the brain is able to "perceive", be aware of, something. Consider, for instance, a variation of the Müller-Lyer illusion, in Fig. 1.1. What arrow shaft is the

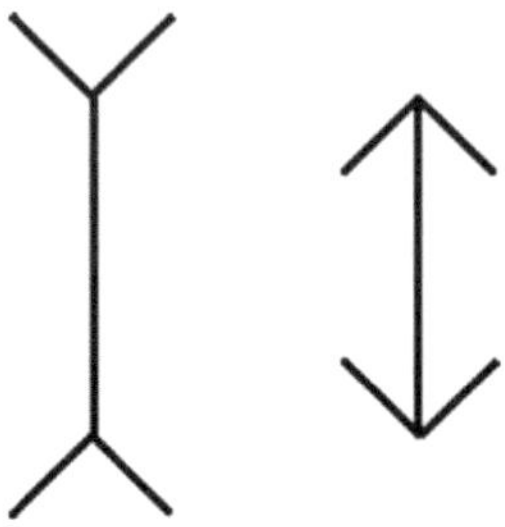

Fig. 1.1 Müller-Lyer illusion

longest? Many people tend to say the one on the left-hand side, with the shorter line segments—the arrowheads, which are called the fins—protruding outwards. In reality, both shafts are of the same length, but the interesting thing for the purpose of this paragraph is that if the fins are subliminal, that is, printed in ink that is barely visible in such a manner that subjects will claim they do not perceive the arrowheads, the illusion still persists. What this phenomenon is telling us is that there is a part of the brain that indeed perceives the arrowheads and can inform other areas of the brain to finally commit an action, in this case, to pronounce the words "that line segment on the left is longer than the other". Are you wondering about the chain of events in your brain that forces your mouth to pronounce those words commanded, or being biased, by a fragment of your brain that is able to perceive the subliminal arrowheads? Could it be that there is more than one self in your brain, one who is consciously aware of things and another (or even several others) that do not participate in that conscious awareness but yet control the "conscious self"? The answer is provided in the next sections of Part I, and Sect. 1.4, in particular, describes in simple terms the basis of brain function and activity, the neuronal chains of events that make us do things.

These phenomena of perception without awareness, unconscious perception if you will, that go under names like subliminal perceptions, blindsight or blindtouch, were pioneered at the end of the nineteenth century by Boris Sidis—in 1898, he published *The Psychology of Suggestion: A research into the subconscious nature of man and society*. The blindsight phenomenon occurs in people who, due to lesions in the part of the brain responsible for vision (the primary visual cortex or area V1), are cortically blind; they cannot see because their brains cannot process visual inputs from the otherwise normal eyes, but yet are able to respond to visual stimuli that they do not consciously see. Thus, some parts of their brains do "see", and there are some findings of what these brain areas are, although in principle, only the visual cortex can correctly interpret visual signals from the retina in the eyes. Similarly, blindtouch, the tactile equivalent to blindsight, is another interesting phenomenon that shows how distinct parts of the brain can process information but may not be necessarily available for a global integration into a conscious percept. Those interested in this topic of perception without awareness and its consequences in our behaviours and are not afraid of technical jargon can consult the book *Out of Mind* (de Gelder et al., eds., Oxford University Press, 2001). Hence, this Müller-Lyer illusion with subliminal arrowheads and in general the world of subliminal stimuli start to indicate that things or events we perceive may not be exactly as they really are.

To further prove the uncertainty of our perceptions, please take a look at Fig. 1.2 and, without reading the figure legend, figure out what you see.

If you are an adult, you will most likely see two people, perhaps lovers, in a jar. But young children tend to see a few dolphins moving about, because they may have never seen naked people embracing, lovers in action. *What you have in memory determines to a large extent what you perceive*. This has been known for a long time in the neurosciences, for example, Friedrich von Hayek (1899–1992)—although he was not a neuroscientist, but an economist—expressed it lucidly as "perception as the source of memory and as the product of memory". That is, what you perceive enters your brain and becomes part of your memory that in turn will determine what you perceive next time.

Not only that, but also *what is in memory and what is expected* determine what you see. You can see in Fig. 1.3 the head of a rabbit. Isn't it? If I had told you (primed you) that the figure will show you the head of a duck, what would you have seen?

And it should not be forgotten that perceptions depend on our sensory organs. Ostensibly we all have the same sensory organs thus we can assume that when we talk to each other we expect the others to have similar perception of reality. However, their internal biases and memories will be different and that will alter such perception as demonstrated above. If you want to know what it feels like not to have some of those senses, you can read Helen Keller's "*The World I Live In*", an account of reality by a deaf and blind woman who described her sensations and the associated workings of her imagination.

Fig. 1.2 Dolphins and people in a jar—it is your choice, or is it not?

Fig. 1.3 Rabbit or duck? Priming will make a difference

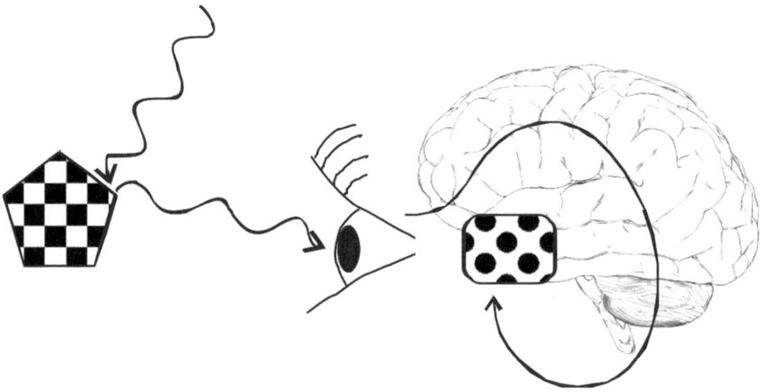

Fig. 1.4 Perceptions (the experience of reality) are transformed by the brain cellular networks where information resides in the form of memory and other subjective tendencies resulting from cravings, prejudices, attachments and other dispositions, so reality is experienced through the veil of the information we have in the brain, such that many times we do not see what we see but rather what we need or want to see. As the philosopher and statesman Francis Bacon (1561–1626) said, "Human understanding is like an irregular mirror, which distorts and discolours the nature of things by mingling its own nature with it"

All these are important insights; it is indicating that perhaps we cannot trust perceptions too much. What we perceive depends in part on what we have inside our brains, as depicted in the cartoon of Fig. 1.4. But if the self is a perception—as defined above, the perception of the unity and continuity in behaviour and cognition—can we then trust it? Mental constructs are therefore abstractions derived from the perceptions, and the perception and identification of patterns is a foremost job of brains. After all, brains emerged to assist humans and other animals in avoiding predators and finding nourishment. The success of each endeavour is predicated on the ability to model the environment and predict future events. This in turn is facilitated by identifying patterns in the events occurring in nature. And naturally, perception of patterns in behaviour and cognition follow suit, and here we have the beginning of the creation of the model of our own mental worlds, that is, of our personal identities.

Hence, if the brain's main job consists in making models of the world and situating us in that world, which results in the foundation of the sense of the personal identity—the famous self—then it seems certain that selves will naturally emerge. In understanding the emergence of the self as an inevitable consequence of having an embodied nervous system (again, by this is meant a nervous system within a body) in a changing environment, it is wise to reflect on how it evolved, how brains and the associated cognition developed. To understand a phenomenon, it is always sensible to go back to its beginnings and explore how it evolved; hence, the following sections depict the scenario where our cognitive power and thus our personal identities or selves emerged.

1.2 The Usefulness of Brains

A good start is to consider why we have a brain. Certainly, brains did not appear for us to philosophise or play cards. The brain appeared because, like other organs, it has adaptive value; it confers on the organism adaptability to the surroundings. Organisms need to respond to a changing environment, and for that, the need arises to situate you (the animal) in the environment. You can now see how the sense of identity, a self, emerged out of the primordial necessity to situate the organism in the environment. Associated with being situated in space (and as well as in time for animals with more complex brains) comes adaptive value: knowing your way around the forest helps find food and avoid becoming food. The processes of sensing and responding to the environment are normally conceptualised by neuroscientists under the general name of *excitability*. This is not the psychological connotation of excitability; rather, it refers to the fact that neurons communicate among themselves using electric currents, changes in voltage. In the very beginning, as organisms developed, from the basic goal of generating coordinated responses to stimuli, there appeared cellular mechanisms that relied upon electrical activity. All neurons constituting animals' nervous systems possess this feature of excitability (see Sect. 1.4 for the very basics of neuronal dynamics) owing to molecules such as voltage-gated ion channels and many other compounds that are studied in neuroscience. These mechanisms exist even in archaebacteria, so this is not a recent invention of nature. Electrical excitability thus appeared long before neurons or nervous systems emerged, and it serves its purpose from sea sponges to mammals. For a very concise view on the origins of the nervous system, interested readers can peruse Greg Miller's "On the Origin of the Nervous System" (Science, 2009, 325:24-26. doi: 10.1126/science.325_24).

This form of excitability favoured organisms to become increasingly adaptable as their responses to external stimuli were faster and more accurate—it is much faster to send a message from the brain to a muscle using electrical activity via neurons than using chemicals, like hormones, via the blood. The sensorimotor transformations performed by nervous systems permitted better control over the environment, which is the main task that brains allow the entities that possess them. The reason we have brains is to sense our surroundings and enact motor actions in response to that sensing, as eloquently expressed by the Nobel Prize Roger Sperry: "Instead of regarding motor activity as being subsidiary, that is, something to carry out, serve and satisfy the demands of the higher centres, we reverse this tendency and look upon the mental activity as only a means to an end, where the end is better regulation of overt response. Cerebration, essentially, serves to bring into motor behaviour additional refinement, increased direction toward distant future goals and greater adaptiveness and survival value. The evolutionary increase in man's capacity for perception, feeling, ideation and the like may be regarded, not so much as an end in itself, as something that has enabled us to behave, to act, more efficiently" (Sperry, 1952).

Hence, it is all about sensing and acting. This is how things started and then evolved little by little, foreshadowing human intellect. We shall see in other sections that the advent of consciousness and self-awareness have adaptive value too. But before, it is advisable to consider some examples that illustrate the, mostly, graded evolution of cognition so that it is understood that our consciousness, including self-awareness and cognitive power in general, did not appear all of a sudden. To inspect that continuum in the evolution of nervous systems and cognition, let us review in the next sections some very interesting observations performed in experiments with animals, from humble invertebrates to nonhuman primates. Their own selfhood will become manifest.

1.3 Cognition Without a Brain: The "Thinking" Slime Mould

A brief interlude now to show that even without a brain, without a nervous system in fact, organisms can display aspects of cognition. Knowing these phenomena can illuminate our understanding of how our own mental competency developed and that of other highly developed animals from non-human primates to corvids and octopuses. This will be explored later. Whereas cognition is usually associated with nervous systems, we have to consider that the notion of cognition (or intelligence) is wide and to a large extent abstract. It is after all a term we humans created to characterise certain aspects of animal behaviour. Much like definitions of consciousness or life itself, these are terms we devised, and their definitions are relative and arbitrary (chapter 7 and section 9.1 contain suggestions to define consciousness and life, respectively),

It is therefore quite interesting to witness how the slime mould, *Physarum polycephalum,* is able to transport nutrients between distant parts of its extended body such that it optimises maximum coverage of nutrients yet expends minimal energy on transportation. All this, naturally, without having an idea of what optimising means or what energy is, as Physarum does not have any nervous system. This humble organism is considered a very minimally cognitive system (Vallverdú et al.,

2018). Its cognitive powers are derived from two aspects that are very familiar to us as they also control our behaviours: reward and punishment. So you see, we are not that far from the slime mould! The main difference is that in our case, we normally use the chemical and biophysical machinery of the brain (that is, for those of us who use it!), but the mould has to use other chemical machinery and biophysics that are not connected to cells like neurons.

We have all seen moulds, that soft green or grey substance that develops on old food or your shower stall. For those who are not familiar with this remarkable creature, this plasmodium is not made up of many cells; it is just one cell that can grow up to many centimetres when conditions are appropriate. This large cell feeds on microscopic particles, and it transports the nutrients along its whole length owing to a protoplasmic network (protoplasm can be considered as the interior of the cell). It manages to cover all available sources of nourishment that it can find (reward) while avoiding repellents (punishment) in such a manner that the transportation inside its cell body is optimised. Its protoplasmic network is optimised for maximum coverage of nutrients yet minimum energy expenditure on transportation of the intracellular material. In a notable—and funny—experiment, the network built up by Physarum to transport nutrients between distant parts of its extended body was found to approximate the Canadian national highway network (Adamatzky & Akl, 2011). This experiment illustrated how sophisticated the networks to transport nutrients between distant parts of its extended body can be. In short, the scientists cultured the plasmodium in agar plates cut in the shape of Canada and placed oat flakes—a delicacy for the plasmodium—in places of the map representing Canadian major urban areas. At the start of the experiment, they put a bit of the slime mould in the "Toronto area". Within 2 to 5 days, the plasmodium spanned the urban areas (where the oats had been placed) with its network of protoplasmic tubes in a manner closely resembling the Canadian engineers' design of the transport network, as represented in Fig. 1.5. Here, we see how an optimal transport network, as the Canadian highway system intends to be, is replicated by the humble plasmodium, and all this is achieved following basic aspects of cognition: sensing the environment

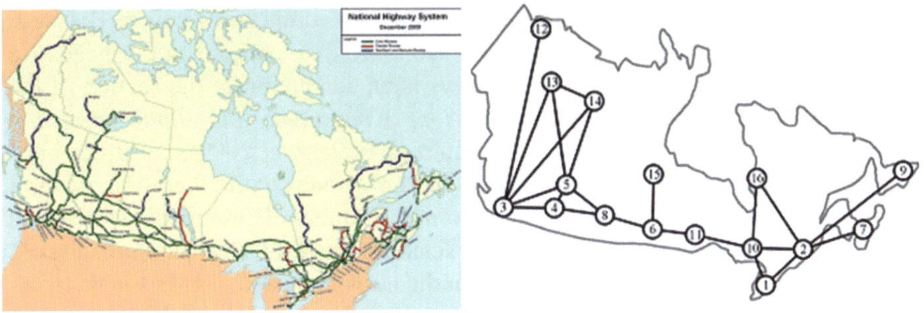

Fig. 1.5 Physarum approximates the Canadian highway network, on the left-hand side. Shown on the right side is an example of the slime mould's covering of the agar plate on which it grows, approximating very nicely the Canadian highways. (Adapted from Adamatzky & Akl, 2011)

(searching for the reward of oats) and acting on it (going for the oats and distributing this reward throughout the whole organism). Further refinements in this perception-action cycle result in our common behaviours, so perchance we are not that special after all.

That is not the only one of Physarum's cognitive accomplishments. It has memory, too. Apparently, it remembers past feeding events, the memories being embedded in the hierarchy of tube diameters inside its body, specifically in the arrangement of thick and thin tubes in the network which makes the organism redirect its movements towards the food (Kramar & Alim, 2021). Now, this is not as amazing as it sounds, if one understands the essence of memory. Memory consists in storing the past in a certain structure. In our case, the structure is the nervous system, in the mould it is the tubes inside the cell, and in metallic alloys it is the metal structure—yes, metals too have memory, they are termed shape-memory alloys and can be trained to move on their own as these structures remember their original shape. The more one understands the essence of things, the less surprised one is about the diversity of phenomena.

Whereas the slime mould performs these cognitive feats without the need for neurons, there is not much more that it can achieve. The emergence of neurons packed in nervous systems allowed organisms to generate fast coordinated responses to stimuli coming from the environment, from avoiding threats to hunting prey. Let us then proceed to see how primitive nervous systems arose and inspect the cognitive powers and self-awareness of a humble worm. But before that, a brief description of the fundamentals of brain activity and function is in order.

1.4 A Brief Tour Through the Brain, for the Non-specialist Tourist

It is now time to review the very basics of how the cellular circuits of the brain operate. We will present the essentials of information processing by the nervous system, at least those fundamentals that will help us comprehend how we behave and how the selves that inhabit each of our brains emerge and compete for power. Because, as you will see in future sections, we do have more than one self in our heads—a tumult of inner voices indeed! What follows is an adaptation of the section of the same title in the book *The Brain-Behaviour Continuum—the subtle transition between sanity and insanity* (Perez Velazquez & Frantseva, 2009), but it has been simplified so that most readers can understand how brains work; therefore, this is a true but very simplified description of neuronal dynamics. Those readers interested in more details can consult any basic neuroscience text or the aforementioned book.

In brief, the essence of the functioning of the nervous system consists in the communication between neurons, one of the two main cell components of the brain. The other cellular elements are called glial cells and are fundamental to provide support for neurons and to modulate their interactions, so a brain would not work without neurons or without glial cells. Neurons communicate by electrical interactions as mentioned in Sect. 1.2. These interactions are called action potentials, or neuronal spikes, and represent changes in voltage (depolarisations) in the neuronal membrane. The depolarisation spreads through the neuronal body, actually through a part of the body called the axon that will ramify and will reach other neurons, causing these neurons, in their turn, to become more depolarised. The contact between neurons is called a synapse, and chemicals, called neurotransmitters, are involved in the synaptic transmission. For brevity, let us skip these details that can be found in elementary textbooks for those keen to know more about the intricacies of neuronal connections. If the aforementioned depolarisation is enough to reach the threshold for a generation of an action potential in one neuron then the chain of activity will proceed to other neurons connected to each other. A neuron will need to receive many action potentials from connected neurons to reach the threshold for firing another action potential. This means that a sort of synchronous arrival of action potentials from several neurons have to arrive at the target neuron. This concept of synchrony of neuronal activity is thus essential for the functioning of the nervous system. If neurons did not synchronise their action potential firings, then the target neurons (also called post-synaptic neurons) would not become active, and the chain of activity would end; this would be bad for our brains. But if too little synchrony would be bad for our brains, too much is equally dysfunctional; for example, epileptic seizures are a manifestation of too much neuronal synchronisation. A general scheme based on this notion of "just-enough-synchrony" for the transition between healthy and unhealthy brain states, including neuropsychiatric syndromes, can be found in the aforementioned book *The Brain-Behaviour Continuum*.

Hence, it is not one neuron that sends an action potential to another and the chain continues; it is more like a mass action. Many neurons fire spikes in synchrony that arrive at another population of linked neurons that in turn become active, generating synchronous volleys of action potentials and spreading the activity down the chain of neuronal connections. This is, in essence and in extremely simple words, the basics of the brain and the whole nervous system functioning.

Now let us take a tour through a normal brain while it is processing information in a conscious state. What is viewed in this trip will depend on what neurophysiological feature is recorded. Let's assume cellular activity is recorded, those aforementioned action potentials. Travelling through the brain cell networks, activity will be noticed in almost all cells, for the neurons (the brain in general) never rest. The trip from neuronal network to network will not reveal much interest, because so much activity is going on all the time in all places. The traveller will be overwhelmed. Now imagine that a relationship between activities of neurons can be measured. For instance, the degree of the synchrony between neuronal networks can be measured, and imagine now that this measure is colour-coded: red for high synchrony and blue for low. Now, the traveller inside the brain will see some cell ensembles appearing in different colours, and some patterns may be discerned. The patterns that the traveller sees are extremely brief, lasting for no longer than 200 or 300 milliseconds. This is the normal brain activity during wakefulness. More specifically, this is the normal activity in the thalamocortical circuit, a connection between the cortex and the thalamus. The brain surface is called the cortex, and Fig. 2.4 shows the four main cortical regions. Visiting other centres deep in the brain such as the brainstem or even deeper in the spinal cord, more uniform and longer-lasting red patterns would be noticed because the synchronisation between neuronal firing is more stable and longer-lasting in these regions than in the brain cortex. An example is a central pattern generator controlling breathing rhythms that needs to keep relatively constant relations among the activity of its neurons. All seems fine so far in the tour. The curious traveller can observe different patterns when the touring happens during slow-wave sleep. Deep sleep, when dreams are absent, would manifest a more uniform red colour throughout main brain areas like the cortex and thalamus because, with the organism at rest, there is not much sensory processing and the neuronal networks tend to become synchronised when deprived of sensory inputs.

If, instead of the biological matter of neural tissue, the tour takes place on a dynamical state space—an abstract state space where neuronal activity is represented using some mathematical tricks derived from equations—then the traveller would now move following trajectories that represent the brain dynamics in this abstract state space. The trip is fast-moving amongst regions of that space that look like black holes. These are attractor-like states, engulfing activity (trajectories) as soon as it gets closer, but the trajectories almost never get so close to being totally engulfed. And if they do become engulfed, it is only temporarily, because noise and fluctuations derived from internal and external sources will push the traveller out of that attractor; the black hole will be left behind, and another will be approached: a fast-moving, non-asymptotic (sorry for the technicality!) journey made up of transient activity (for readers interested in the so-called chaos theory, more formally dynamical system theory, we inform them that this field of physics has had and is still having important applications to neuroscience, and some basic technical details about this formalism applied to brain research are expounded in Chap. 2 of the book *The Brain-Behaviour Continuum*). Again, all seems fine in this brain.

But the tour during the awake state through other brains that display some pathologies will reveal to the traveller some unusual colour patterns. For instance, the excursion in the neural tissue could reveal more uniformity in the colours that represent the synchrony, lasting longer than milliseconds. These patterns may be sustained for seconds or minutes, which in "brain time" is long. If the intrepid traveller encounters this scenario in brain regions such as the striatum and thalamocortical circuits, this is a hint of this brain belonging to an individual with a motor disorder like Parkinson's disease. If this pattern is seen in hippocampal and temporal networks, then this brain belongs to an epileptic patient. The presence of this pattern in the auditory sensory cortices may indicate that this brain's owner experiences auditory hallucinations, as in schizophrenia. The uniform colours may be seen in some very specific brain regions in a person with depression, ruminating endlessly on the same thoughts. The dynamical trip through each of these brains would be also different from the previous. Now, the traveller would be stuck in one of those "black holes", the attractors in brain activity, for longer durations, at times wondering whether there is a way out. It would take large perturbations (medications or electrical brain stimulation) to abandon this attractor-like state. Unsuccessful treatment will bring the traveller back to this attractor after some time. These brains are different from the normally functioning brain and result in altered output or behaviour.

A final consideration about a most basic feature of brains: the neuronal networks tend to be reciprocally connected. In other words, there is recurrence of activity. Recurrent cellular ensembles and their close relative, self-referentiality, are one of the pillars of biological phenomena (Goldenfeld & Woese, 2011) and consciousness in particular. For those interested, there is an extensive literature on this topic of recurrence that goes as well under other denominations (re-entrance, reverberating circuits, bottom-up and top-down neural activity, etc.) that, basically, denote the same aspect of recurrent, feedback and feed-forward, activity between reciprocally connected cells. This property of neuronal circuitries is present already in primitive systems, exemplified in the organisation of structures that appeared very early in evolution such as the central pattern generators (like the aforesaid one that controls our breathing rhythm). This feature is also the reason why brain activity recorded as an electroencephalogram displays oscillations of the brain waveforms, which denote a healthy brain as opposed to an almost flat line without apparent oscillations in coma and other pathological brain states (but beware that pathological neural oscillations do exist, namely, in Parkinson's disease or in epilepsies).

And one more crucial consequence of recurrent activity: it favours the maintenance of temporally sustained patterns of cellular activity. In this manner, it is hard to establish a beginning or an end in a chain of neuronal discharges. We tend to think that upon receiving a sensory stimulus (e.g. seeing a wine glass), the chain of activity in our brains starts, and finally, it may end up in us pouring wine in that glass and drinking it. It is not that simple, recall the comments above about how what we have in memory determines what we do: the image of that glass we receive via our eyes forms in parts of the brain that process visual inputs and those brain areas interact with the activity present in other brain regions where images of wine are stored; therefore, the visual glass input is just a brief spark that ignites what was already there, so to speak. Furthermore, the final neuronal action is not in those neurons that activate our hands' muscles to pour the wine (either a red or a white wine would have been previously selected by neuronal networks in other brain regions, and this relates to the topic of free of will that shall be discussed in Sect. 2.2), because that motion creates a feedback proprioceptive stimulus that is sent back to the brain and there the chain of activity continues. Gilles Laurent advised us to change our perspective on this matter of finding the beginning and the end of neuronal activity: "Our thinking generally ignores the fact that, with the exception of motor neurons, a given neuron is never an end-point or its response an end-product […] Thinking about sensory integration in these active terms (considering 'responses' not only as products but also as ongoing transformations towards some other goal) may be helpful […] to understand some brain operations" (Laurent, 2002). Note he says, "with the exception of motor neurons", as if these were the end-point or end-product, but as it was just mentioned above, these motor neurons by virtue of moving muscles cause a proprioceptive feedback to the brain that continues the chain of neural activity. Hence, as it is becoming apparent, there is almost no clear end and no clear beginning to any neuronal chain of activity that determines this or that behaviour; activity in one chain becomes another activity in other chains. This is the incessant 'being and becoming' in neural tissue with implications for self and conscious awareness as it will become apparent in the following sections.

Incidentally. we have all experienced and witnessed results of these incessant chains of neural actions and the associated sustained patterns of brain activity. For example when we perform a series of routine actions that are nearly identical time after time, or when we answer with almost the same words similar questions we are asked, or during the typical "mind walking or wandering" when we go from one thought to another related almost without any conscious awareness of this mental walk (this automatic wandering of mental activity from thought to thought may be cumbersome in some occasions and may lead to troublesome over-elaboration of events that occur to us, but there is remedy for this, and in Part III, we will describe some therapies to protect us against this almost unconscious mind wandering). And if you have rodent pets or work with rodents in laboratories, you will have probably seen the stereotypical grooming behaviour in rats when these animals follow a standard set of motions when they groom themselves, another manifestation of the chain of activations of neuronal networks.

An additional consequence of these recurrent cell circuitries, that is important in order to make sense of some observations that will be exposed in following sections to shed light on the nature of the self and consciousness, is that whereas brains have areas that seem to be anatomically and functionally independent to process specific stimuli and sensorimotor transformations in general —the auditory cortex for hearing, sensory cortex for touch, motor cortex for movements, etc.— in truth such segregation is not at all strict. For example, the visual cortex sitting at the back of the brain receives feedback input for frontal cortices (at the front of the brain as the name indicates); basically, in the brain everything is connected to everything else, partitions are nothing but fuzzy, which sometimes may lead to pathologies. Synesthesia is a fascinating example of this imperfect partitioning of brain regions. In synesthesia one sense, for instance hearing, is simultaneously perceived as if by another sense such as sight. The Piano Man, Billy Joel, has described his ballads in shades of blue and green, while his rock music is red, orange or gold[2]. Synesthesia is also an example of different from normal brain functioning having a "positive" output (meaning something that does not perturb too much the individual who can have an almost normal life). Hence, it is useful to always have in mind that due to the abundant interconnectedness among brain cell networks, basically the whole brain is a vast connected system where one spike (action potential) could travel from one region to any other. Virtually all brain functions involve routing neuronal activity among a widely distributed network of areas. Thus the importance of the concept of coordination in the activity of brain cell circuits, which is the subject of a field of study called brain coordination dynamics.

[2] www.psychologytoday.com/ca/blog/sensorium/201205/hes-got-way-about-him

1.5 The Cognitive Powers of a Worm and its Extremely Primeval Selfhood

If the evolution of cognition and consciousness has been a continuum, a long evolutionary progress that precludes all-or-none conclusions about some animal having cognition, then, do other animals have a sense of self? Or is selfhood an exclusive Homo Sapiens phenomenon? It is clear to whoever has had pets, especially dogs and cats, that these animals have a perception of their own identity. They can situate themselves in an environment, be it the house, the garden or the street, and they are able to anticipate what things may occur in that environment with them as witnesses, that is, as separate individuals from the others that may be causing the, for them, interesting or dangerous things. As we will see below, many animals pass the "mirror test", an experiment that indicates these animals possess self-recognition. Before we explore the complex multi-cellular animals, let us reconsider other creatures, simpler entities. What of primitive animals? Do they have a sense of identity?

Let us examine some invertebrates, the earthworm, or a snail, both creatures being common sights in a garden. Watching the crawling of one of these specimens you may have noticed that as it moves it is not perturbed by the pebbles or twigs it encounters. If, however, you touch it, it will withdraw, retract. The withdrawal reflex in worms is a good illustration of a primitive sense of self. These animals respond to cutaneous touch but not to self-induced touch. They therefore differentiate between information about the outside world and the self-produced sensory input. How can they differentiate the source of these stimuli if these animals have a minuscule brain? Does size matter in this context? Technically, the set of a few hundred neurons that form their brains may not even be considered a brain, rather an interconnected mesh of neurons (but, isn't it what our brains are?). Nonetheless, that is sufficient to display the phenomenon known as the corollary discharge, the reason why they differentiate self-produced inputs from others-produced alterations in their path. Can this corollary discharge be considered cognition or self-awareness? Before launching into the muddy waters of semantics or definitions of cognition and consciousness, let's inspect for a brief moment what this corollary discharge is and its significance in the human brain.

The corollary discharge, once termed Feinberg theory (Feinberg, 1978), consists in the transmission of a copy of motor commands to the brain sensory cortex where the expected sensation that will be caused by that movement is generated. For those new to neuroscience, the brain cortex is comprised of diverse areas. Some are motor areas (motor cortex) that produce movement. Neurons in the motor cortex send their action potentials to the spinal cord, then the signal travels to the targeted muscle and produces the desired movement. Other regions are sensory cortices that process sensory input like vision or touch. Figure 2.4 depicts the main cortical lobes and regions. Some regions called association areas act as coordinating centres, receiving input from multiple regions and forming functional connections between them. The parietal and prefrontal cortices (to be explored in the next sections) receive inputs from multiple regions and therefore form connections between sensory and motor areas, integrating incoming sensory information. Going back to the corollary phenomenon, this discharge does not produce any movement itself but instead is directed to regions of the brain to inform them of the impending movement. As a result this phenomenon allows for a primordial distinction of self and non-self, because the worm, or you, can anticipate what you will feel when advancing to touch a pebble, but if the stone suddenly comes at you then the anticipation fails and this sort of mismatch between expectations and reality makes neurons in certain brain regions (the aforesaid motor cortex) to produce actions: if you are a worm a retraction is produced, if you are human... Well, if your discharge is dysfunctional you may present some symptoms of schizophrenia like hallucinations, the reason being is that you will fail to differentiate between sensory impact arising from your own activity versus changes coming from outside, and naturally this could lead to not only hallucinations but many challenges in assessing the world (about the association of this phenomenon with schizophrenia there is still research going on, a recent review is Parlikar et al., 2019). On the other hand if your corollary discharge works fine you will have problems tickling yourself (yes, it explains things like why it is very hard to tickle ourselves). Hence the corollary discharge seems to be a basic feature of nervous systems that is found throughout the animal kingdom, from worm to human. It can be considered as a general mechanism of the brain's internal monitoring, which in the case of movement keeps track of what is about to happen as a consequence of our own movement.

But this phenomenon, while allowing for a primordial distinction between self and others, does not imply the kind of self-awareness associated with more advanced animals, including humans. The few cell numbers and coupled neural networks in worms' nervous systems do not allow these animals, presumably, for a self-conscious determination of the withdrawal, being just a reflex. How could our worm have a more advanced self-recognition? Is it conceivable that the addition of more neurons, more recurrent neural ensembles between the sensory (perception) and motor (action) systems will result in features of conscious awareness emerging? This possibility will be dealt with in chap. 6 in an effort to understand how we move from the typical reflex arc to self-monitoring of our actions. Let us say for now that the quantitative changes in the brain (such as more cells) result in qualitative new cognitive capacities.

Nevertheless one may ask, how do we know that worms do not possess self-awareness similar to ours? After all, we cannot communicate with them. While scientists should always be cautious when asserting certain ideas, yet I believe we would (almost) all agree that it is extremely probable that these animals do not share some particular features of what we term consciousness, especially what could be called the higher order features, among them a deep and clear self-awareness. Now that we have encountered the issue of consciousness and how to assess or define it, it is time we start demystifying it, demystification that will be settled in chap. 7. For the time being we can just say that what we term consciousness can be defined by several properties (in more detail in chap. 7) including some primordial, like processing of sensory stimuli and responding to changes in the environment, and other higher features in this presumed hierarchy such as self-recognition. Consciousness is thus a constellation of features. And because even the simplest of creatures can associate conditions with responses and consequences, such that harmful ones are avoided and beneficial responses reproduced, it could be fair to attribute some consciousness to all organisms. Hence, from this perspective, simple organisms possess some features of consciousness, so one could say they are conscious to some extent, but again, only the highly advanced animals may be said to possess the full list of features of consciousness. In this manner, the so much debated question —many times in esoteric terms— whether everything is conscious or not, can be partly settled. A bacterium that can sense and respond to the environment is somewhat conscious and hence those against the idea of pervading consciousness will take refuge in the notion that bacteria are not fully conscious because they lack many higher order features, whereas those in favour of panpsychism will be happy to see that even bacteria display some consciousness attributes, thus everybody is satisfied.

But while discussing how one can attribute features of consciousness to other creatures, or even to other humans, we encounter another issue; what is it that one perceives from others and how does this influence our attribution of cognitive features to other animals?

1.6 On the Essence of Selves: Properties All We Perceive

When considering selfhood, the essence of that "I", we must consider that we do not have access to the essence of things, only to their properties. A plant is green because we see the green wavelength of light reflected. The "greenness" is a property that our vision perceives. When physicists "see" atoms or elementary particles, they only see their traces in a cloud chamber—a particle detector where particles interacting with gas molecules result in visible trails of ionised gas particles; hence, they see properties of the atoms but not the particles themselves. It is the same with people. We experience their (psychological or physical) properties when they behave and as well those of their skins when we touch them or look at them, but knowing the true essence of one person is far from trivial and almost unreachable. Once again, D. Hume said something very pertinent to these points: "What we call a mind is nothing but a heap or collection of different perceptions, united together by certain relations and supposed, though falsely, to be endowed with a perfect identity".

But while the essence does not, or may not, change, properties do change depending on what/who is in front. Consider weight, a property of a mass, of a piece of matter. Weight is different on the Moon and on Earth; it depends on what that mass has "in front" ("below" would be more appropriate in this case), or communicates with—we can take gravity as a sort of "communication", if you will, between two particles. And similarly occurs with people: they tend to adjust their behaviour—psychological properties—depending on who is in front. The essence of that mass, say the number of atoms that constitute that piece of matter, on the Moon is the same as on Earth: essence remains the same, properties change, and we only have access to properties. We encounter again words by von Hayek, this time on the topic of what we perceive and how we (our brains do it automatically) establish relations among those perceived features: "All we can perceive of external events are therefore only such properties of these events as they possess as members of classes which have been formed by past linkages. The qualities which we attribute to the experienced objects are strictly speaking not properties of those objects at all, but a set of relations by which our nervous system classifies them" (von Hayek, 1952). But it should not be too frustrating that the true essence of natural phenomena remains obscure to our intellect, for as the physicist Niels Bohr said, "In our description of nature the purpose is not to disclose the real essence of phenomena, but only to track down as far as possible relations between the multifold aspects of our experience" (Bohr, 1934); hence, we can be very busy indeed establishing those relations among things, which in part is the job of scientists. And if this is not enough encouragement, consider what another physicist/mathematician advised: "The aim of science is not things themselves, as the dogmatists in their simplicity imagine, but the relations among things; outside these relations, there is no reality knowable" (Poincaré, 1917); so you see, there are those who think our current cognitive powers cannot grasp much more than relations among properties, and to reiterate again, forming relations among things is done automatically by our brains because we have to remember that brains evolved to make models of the environment and predict, anticipate, events. How the brains make models of our own psyche—that is, a model of the organ itself, and how about this for self-referentiality!—will be detailed in future sections describing neuropsychiatric syndromes, as this is fundamental to comprehend how we create our self-model.

Therefore, trying to perceive the essence of the self may be a futile search. It is already difficult enough to understand oneself, now try to understand others. And yet we constantly make judgements about others, we think we can assess the feelings, cravings and hesitations of others. Why do we think that we are entitled to such views? The philosopher Thomas Nagel once asked: what is it like to be a bat? This is in fact the title of one of his essays ("What Is It Like to Be a Bat?" The Philosophical Review, October 1974) where he inquires how we perceive others and especially how one decides to map onto other beings' behavioural or mental actions. Note that he is not asking what it is like for you to imagine a bat's experiences, to put yourself onto the bat's mentality using your own mental constructs, feeling the bat's world filtered through your human perspective; rather he is asking what it would feel if you were that bat, in other words, you would have be that animal itself, experiencing the bat from its own viewpoint without passing it through your human veil. Such a feat seems unfeasible as you would have to cease to be yourself, the product of your own brain. "You" would have to become the product of a bat brain. Does it become easier if instead of a bat we talk about another person? If we have access only to the properties of the person in front, you perhaps can imagine what the other person is experiencing, perhaps empathising to a large degree, but in the end you can never be sure, for this is a mapping of your own thoughts, your own psychological traits, onto your companion. Following this reasoning, I could concoct a definition of myself as: to you, I am the properties of a self-referential system that attends to the name of Jose Luis —self-referential system because our cognitive machinery endows us with self-awareness, self-monitoring of our behaviour. If we consider that, as aforesaid, the sense of the self is a perception, that some parts of our brain can perceive properties, and that other sections of the brain (presumably integrative areas such as parietal and frontal cortices) yoke those perceptions into a unified construct as will be detailed in chap. 3, well, you have the start for a sense of self.

The moral of the story of the true essence versus properties is that care should be taken when judging others, when assessing the cognitive faculties of people and those of animals that will be explored in the next section, where questions about intelligence or self-awareness of these creatures will be pondered. Along these lines, it is always a refreshing exercise to go out in the street while trying to abolish all the preconceived notions, prejudices and similar thoughts which bias our views, because it is almost a reflex that when we encounter another person in front thoughts start to arise in our minds about a variety of aspects this person may or may not have. Judging others is an international trait, these opinions about others being another enslaving mental state that is very hard to avoid due to that reflex-like nature, but if you can avoid it, you will have progressed in your path to liberation. Very much related to this reflex-like habit of judging others by their appearances is the matter of racism, about which some words are devoted in Sect. 3.1.

One thing that can be asked now is whether the essence of self, of the personal identity remains stable or changes too. It was suggested in the first paragraph above that the essence of things does not change, only the properties. Remembering the words of the Brazilian author Paulo Coelho in his novel 'O Alquimista', "the dunes change with the wind, but the desert remains being the same", it is tempting to assert that the essence of our personal identity is stable throughout the lifetime, and only properties —psychological traits, longings, anxieties— change. The remarkable experiences of people with brain injury that will be presented in parts of this book may answer the question, where we will see the sheer obstinacy, the tremendous stability of the illusion of the personal identity; but for now, as we have no access to essences, perhaps it is best not to conclude anything on the matter. What is known with certainty is that our properties change and that, if we are to live long and happy, it is advisable to acknowledge and understand those alterations; being and becoming, realise what we are and in what we become is a first step for a serene life.

1.7 The Personal Identity of Animals: A Self Through the Looking Glass

In the previous section we have seen that it is not a trivial fact to declare an entity with which we cannot communicate as possessing certain cognitive powers as we cannot really feel what that entity is thinking, sensing or feeling, nevertheless in this section evidence will be presented that reveals that some animals possess self-awareness, particularly thanks to the method of the mirror test.

If not self-awareness, there are aspects of cognition present in ancient animals, so before we subject other animals to the mirror test and understand how they recognize themselves, let us briefly examine some illustrations of cognitive powers in animals. Because, as Charles Darwin pointed out, "The difference in mind between man and the higher animals, great as it is, certainly is one of degree and not of kind". Bees, for instance, have emotion-like reactions (Baracchi et al., 2017) which while these may be referred to as "appetitive motivation", in the final analysis these behaviours represent the continuum in cognition; the essentials of cognition are already present in these simple animals, granted by their primordial nervous systems. Like the worm, insects' brains have very few cells as compared to mammalian brains. However, they are also composed of different regions that, just like in our brains, contain densely interconnected cell circuits (recall the explanations about neuronal recurrence and interconnectivity in Sect. 1.4). Some parts are conserved across all insect species. One such region is the central complex, a multi-modal information processing network (Boyan & Reichert, 2011). For all of their intricacies, these primitive brains have limited ability for actions, and although we can endow these creatures with some cognitive faculties as they possess some features of consciousness —recall how our previously introduced worm was able to sense and react to the environment, and sensing and reacting are two main features of cognition— they probably do not possess the higher level features of consciousness/cognition such as self-awareness (the word "probably" in this sentence is just for cautionary purposes, for nobody has reported any mirror test on these animals to our knowledge, but again, it is not easy to be totally, completely, without the shadow of a doubt, sure that these small organisms do not have a sense of identity —well, one study has been done where ants were reported to have passed the mirror test, keep reading). However, to be fair to these animals like insects that live in big communities —colonies— it should be understood that their intelligence, or cognition in general, is so far away from ours that we cannot recognise it, for it is true these social animals, like bees or ants, possess language with which they communicate relevant things for their survival and show elaborate group integration features in their colonies such as division of labour. One should always be careful with the tendency to anthropomorphise everything. As an aside, language is tremendously important for high level cognition to develop, and interested readers may consult a wide variety of studies on the topic. Almost all creatures have a language, perhaps not verbal, but language it is nonetheless: ants communicate via their antennae and chemicals like pheromones. From the continuum perspective, these pheromones giving rise to "mental constructs" —very primitive or simple, but again we may be judging too fast— in the ant's nervous system are foreshadowing more complex mental objects that emerge in more advanced animals as a result of other types of languages and complex brains. Hence, language, and the bipedal posture mentioned below in chap. 6, are two main aspects that promote the progress of high cognition.

While it is among the vertebrates that high levels of cognition can be seen, it would be unfair to close the comments on invertebrate cognition without mentioning the cephalopods (*Octopus vulgaris*), regarded as the most intelligent of the invertebrates. Octopuses (note, the plural is not 'octopi' as sometimes is used because the termination "i" occurs in Latin words and octopus is a Greek word) perform very well in learning and memory paradigms and are comparable in sophistication to some vertebrates. For example, they are able to find rewards in mazes, retrieve objects from sealed bottles and discriminate between objects based on different characteristics. They also recognise conspecifics (Tricarico et al., 2011), which suggests that these animals may have some aspect of self-awareness, but alas, these animals have not been shown to pass the mirror test. Of note, octopuses' nervous system organisation is totally different from that of vertebrates, and they have a pseudoautonomous control of locomotor activity that, apparently, resides not in the central ganglia (akin to their brains) but in neural networks in the tentacles, so their limbs seem to have each one its own "intelligence", as it were.

It is among the vertebrates that cognition becomes as complex as to develop self-recognition. Examples of high intelligence are numerous. Students of avian cognition know very well that corvid birds and parrots are regarded as the most intelligent birds, and their intelligence has been considered as being equal to that of primates. To wit, ravens adjust their behaviour to social context, have extraordinary memory (they will remember if you were good or bad to them), use tools (see Fig. 1.6), display vocal learning and have enormous behavioural flexibility. Magpies are smart too, and they have passed the famous mirror mark test. Not surprisingly, the brain electrical activity (normally recorded as electroencephalograms) patterns of birds are very similar to those recorded in mammalian brains during wakefulness and sleep.

And of course, when we arrive at primates, we find a very advanced cognition or intelligence. Before describing the mirror test and the results, and to close this section on animal intelligence, let us emphasise once again that our intelligence/cognition concept is very anthropomorphic. We consider people intelligent if they do

Fig. 1.6 A corvid using a branch as a tool to find food. Crows have been seen "manufacturing" (wrong word because they do not have hands) hooked-wire tools to retrieve food from holes as this specimen here shows us.

things like mathematics, philosophy, anticipating future events or finding creative solutions for problems. This human narrow definition of intelligence can lead to judgement and discrimination. One can wonder what, say, a chimpanzee's or a crow's concept of intelligence may be, perhaps living in nature for several weeks without any shelter or technological stuff ready to be used for survival, or being able to sense at a distance the presence of predators and preys. The crow and the chimpanzee may possess another intelligence that allows them to survive in the natural elements without the benefit of technology. So next time you emit an assessment or judgement of someone's, human or other species, intelligence, please consider this. Particularly in the Western civilisation, we value intelligence in part to satisfy our need to differentiate ourselves from other animals. Yet, how are we different from the slime mould whom we first encountered? (chap. 10 will deal with this matter on what makes us so special, or maybe unspecial). Remember that the primary drive of all animals is to derive energy from food, whatever that food may be. Despite the simplicity of single-celled and simpler animals' nervous systems, their brains are sufficient to make energy acquisition successful. Humans are complex animals, who required complex brains to find food in a complex environment. We require machines to grow and harvest food and manage farm animals. We require jobs to earn money to exchange for food. In short, we have created a most complicated world, courtesy of our intellects.

Readers interested in more details can consult numerous reviews on animal cognition (e.g. 'Cognition, Evolution, and Behavior' by S. J. Shettleworth, Oxford University Press), there is even a journal by that name (*Animal Cognition*, Springer) where papers on the matter are published. As well, there are a number of technical reviews on the evolution of the brain and intelligence (Cairó, 2011; Roth & Dicke, 2005). Having revised some basic observations that illustrate animal cognition, now the question arises as to whether that cognitive competence allows animals to develop a sense of identity, a self.

1.7.1 The Mirror Test

In 1970, the psychologist Gordon Gallup Jr. published a landmark paper describing a method to test whether an animal possesses self-recognition (Gallup Jr., 1970) or, more precisely, to determine whether an animal has the ability of visual self-recognition. This method was termed the mirror test, also known as the mirror self-recognition test or the mark test.

The experiment goes as follows: an animal, normally under anaesthesia, is marked (painted, or a sticker is attached) on an area of the body the animal cannot see directly but can be seen in front of a mirror. Once awake, the animal is given access to a mirror. If the animal then touches, examines or tries to remove the mark, it is an indication that the animal perceives the reflected image as itself, rather than that of another animal. Figure 1.7 depicts monkeys passing the test, and a funny video where orangutans can be seen passing the test can be found in videos like the

Fig. 1.7 Chimpanzees performing the mirror test (image in the public domain)

one mentioned in the footnote.[3] Primates, including bonobos (it is equivocal in gorillas, at least at the time of this writing) have not been the only animals that have passed the test. In the link indicated in footnote 4, you can see magpies being tested.[4] Dolphins and Asian elephants have been reported to pass the test too. You can watch here the behaviour of a dolphin in front of a one-way mirror[5] and decide whether or not the animal is having fun with itself, which would indicate it has recognised itself in the reflection. Other animals that may have passed it include orcas and even a tiny marine fish, the cleaner wrasse. Keep in mind it is not a trivial experiment, for how can you be sure a fish is trying to look at the mark in the mirror? One has to judge carefully whether the movements of the animals in front of the mirror are unique, unusual or consistent with behaviour associated with self-awareness.

Humans naturally pass the test they devised, but it takes a while! If you have young children, you may want to see whether they pass the mirror test. Normally, children under 15 months of age do not pass it. And whatever your results, please remember this has nothing to do with intelligence; if you have a 12-month-old who passes the test, it does not mean your child will become a genius. Does this mean that self-awareness develops later, that we do not have a self at birth? And if so,

[3] www.youtube.com/watch?v=Okmkn30D0NU

[4] www.youtube.com/watch?v=HRVGA9zxXzk

[5] www.youtube.com/watch?time_continue=4&v=HKa8DBUxP5I&feature=emb_logo

what is the nature of that early personal identity? And how come it grows to such a tremendous extent in adults? More on this later, now let's continue with other fauna.

As aforementioned, there has been a study suggesting that ants are capable of self-recognition (Cammaerts Tricot & Cammaerts, 2015). Some ants were painted with blue dots on their face and, apparently, would groom and appear to try to remove the markings in front of mirrors. Very young ants and others with brown dots, a colour that blended in with the colour of their bodies, did not clean themselves. Interestingly, other ants responded aggressively to those with blue-marked faces, presumably because the colour difference caused them to think the "blue ants" were outsiders, so the fact the marked ants in front of the mirror did not attack themselves seems to suggest that they, perchance, recognised themselves in the mirror. Or not? Now, the scientist in me comes out and starts to be very cautious. As was said above, it is not a trivial task to judge when the animal is doing some uncommon, odd movements in front of the mirror (presumably to reach the marked spot and remove it) or altogether to interpret the animal's behaviour. For starters, ants have compound eyes made from numerous tiny lenses, which are good for acute movement detection but do not offer high-resolution images. Hence, can they clearly see the mark on their faces in the reflection? And even in the affirmative case, why an ant should try to clean itself? The experimenters reported that all (unmarked) ants behaved unusually when placed in front of the mirrors, moving their heads and antennae about rapidly, so one wonders how easy it is to be certain the blue-marked ants were performing those observed movements to clean the blue spot or just that they were agitated, like those unmarked, about seeing another—their reflection—in front. Therefore, while the study was very carefully done, the complications with this type of experiments are such that we should take the results with a grain of salt.

And because nothing is perfect, the mirror test has inadequacies too. Implicit is the idea animals will care about the mark, but as some readers will have thought by now, the possibility that some animals will not care about the mark enough to try to reach it and explore it will yield a negative result that may be erroneous. Related to this point, the test may be relative to the species; for instance, there are animals that do not rely so much on vision, rather on other senses like smell. For example, dogs have not passed the mirror test (at the time of this writing) but have passed the "scent" test in that they distinguish between their own odour and that of other dogs (Horowitz, 2017), a sort of sniff test if you will. In any case, because it is difficult to be certain of what goes on in a dog's mind (or in anybody's mind for that matter), these results can only suggest the possibility of self-recognition in these animals. There is another piece of evidence, not related to mirror tests, suggesting that dogs have body awareness through the understanding of the consequences of their own actions (Lenkei et al., 2021). In this study, the authors employed a similar experimental setup that Jean Piaget—a pioneer in the study of the origin of intelligence in children and whom we will meet in a section below—used while assessing the development of cognition in his own children in that the dog had to recognise that its body was an obstacle to achieve a certain task. Piaget's toddlers were able to comprehend this after the age of about 18–24 months; therefore, dogs seem to reach, at least, one of Piaget's stages of the development of children's intelligence that will be described in Sect. 1.8.2.

Anyhow, the point is that for some creatures, the mirror test may not be relevant at all. Indeed, false negatives (i.e. an animal that is self-aware, sees the dot, and just doesn't care) have been reported in the literature. And there are those other animals that may not pass the test, that is, do not recognise themselves in the mirror, but still have interesting reactions to their reflections. To emphasise it again, it is a difficult experiment to interpret, and only in the cases that are very clear the result should be declared a positive one; the rest should not be judged, neither positive nor negative, but uncertain.

Nevertheless, the mirror test experiments have demonstrated that some animals clearly recognise themselves as individuals, that they have a sense of selfhood. Considering this evidence plus other observations such as that animals are able to understand their situation in space and in their societies, we can conclude that animals perceive the world in relation to themselves. Together with other samples of cognitive powers related to this aspect of understanding that one individual is different from the other (e.g. primates understand third-party social relations), all these are hints that they may have developed selfhood, a personal identity. For instance, many animals live in groups with a strict hierarchy, which demonstrates they are very well aware of their situation in their group, in their social scale so to speak. To wit, wolves live in packs with a strict hierarchy such that the alpha male and female will not allow omega wolves to mate, so they are very well aware of their place within that hierarchy, their social standing[6]. But can their selves grow big, or inflate with self-importance? Are there egocentric maniacs in the animal kingdom, to a similar extent as some humans can be? As anybody who had pets can attest, animals tend to be very egoistic, e.g. they will try to be the first to reach the location of appetising food, they will struggle for the owner's attention, and there are other interesting behaviours pet owners can observe and perhaps enjoy. However, this selfishness does not guarantee they develop egos to the extent humans do. For one thing, they lack the intense social input characteristic of most human societies (for sure the Western societies) with regard to the glorification of the individual, praising that starts early in childhood in attempts to convince the child he/she is a most worthy individual—not that this is totally wrong, but in Western societies there has been lately a surge in attitudes like, in the words of a colleague of mine, praising every poop and breath of the child as worthy of celebration, and as we shall see in Part III, big selves may mean big problems.

And just to make it clear, the apparent examples of altruism of some animals like ants or bees working to help the queen have daughters, or the bees attacking an intruder and in so doing dying (unlike wasps, when bees sting they leave behind part of the abdomen; hence, they are kamikaze animals), or the deer—normally the mother—who runs in front of the wolves to attract their attention and move them away from her offspring, are not that altruistic as supposed. If you were to perform

[6] www.livingwithwolves.org.

a calculation based on genes, you would find out the remarkable fact that in Hymenoptera like bees or ants, the worker sisters are more closely related to their sisters than to their own offspring (hypothetical offspring because workers do not reproduce, but if they were to have babies, they would share fewer genes with them than they share with their sisters produced by the queen, contrary to what happens with mammals where the parents share more genes, on average, with the offspring than with the siblings), therefore helping the queen produce sisters is in the workers' genetic self-interest. For all these very intriguing facts about presumed animal altruism and its explanation based on the "purpose" of the genes, the reading of texts like *The Selfish Gene* by Richard Dawkins is recommended. For humans, as for other animals, selfishness may be necessary for survival. In Viktor Frankl's book, *Man's Search for Meaning*, he writes of concentration camp survivors: "...the best of us did not return". In extreme circumstances and hardship, perhaps altruism is not the best strategy.

These behaviours, whether egoistic or not, occur because our bodies and brains are the product of genes. And this is precisely where the genes may have erred, in the creation of advanced brains—but please do not take 'create' literally, as genes do not have any idea what they do; in fact, there is no purpose in nature, as expounded in Part II. With big brains and high intelligence, now, the organisms, namely us, can go against the will of the genes which seek maximal reproduction, and indeed, we are already doing things the genes would never approve, like the use of contraceptives. The future will tell whether genes made a mistake "creating" the brain (much more on this in chap. 10).

Now that the existence of selves in the animal kingdom has been presented, the question remains as to what type of self the animals experience. Is it similar to the way we humans perceive our personal identities? And what type of consciousness do they experience? Is it possible to even barely grasp these experiences of animals, considering the comments in Sect. 1.6 about the difficulties of mapping our intellects into other entities? Let us explore the possibility of similarities between the selves in animals and the developing selves in human babies that will help us comprehend what our sense of individual identity is doing to us, where it is taking us. The next section reveals some basic aspects of the development of self-awareness in children.

1.8 Of Animals and Babies: The Development of Self-Awareness in Humans

The intention of the previous section was to present the view that there has been a graded development, a continuum so to speak, of cognition and consciousness, so that we understand that human consciousness and cognitive power did not appear all of a sudden but rather emerged from this continuum. In the final analysis, all organisms, from protozoa to humans, share a similar purpose in life: to generate (adaptive) responses to stimuli. Thus, from the simple reflexes or taxes (no, we are not talking about those compulsory and many times upsetting contributions to the state, rather taxes in biology are the responses of organisms to stimuli by automatically moving directly toward or away from the source of the stimulus) to fully conscious and voluntary actions, the purpose is to satisfy the needs for survival—recall R. Sperry's words in Sect. 1.2. It cannot be underestimated that cognition is the result of a long evolutionary process, not considering this will result in an incomplete understanding of cognition and the related aspects of consciousness and self-awareness.

The mirror test presented in the previous section indicated some animals have some sort of selfhood as a feature of their consciousness. Can we fathom what kind of selfhood? Some scientists, like Gerald Edelman, proposed that there are different types of consciousness (we will have to wait until chap. 7 for the demystification of consciousness): primary consciousness, the state of being mentally aware of one's surroundings, and higher consciousness, the recognition of our own acts (Edelman, 2001). Each of these two types of consciousness carries with it, or implies, some experience of personal identity, for you need to be aware of your own self to understand you are surrounded by other things and entities (primary consciousness) and to see you as an active agent of things happening around you (higher consciousness). It is fundamental to consider, when addressing questions about babies' or non-human animals' consciousness, that the assessment of consciousness normally is associated with language: a verbal report is sought from the subject. But in very young babies and other nonverbal animals, there is no chance of obtaining verbal, subjective reports; hence, a different framework for research that extends testing methods beyond subjective reports seems necessary (some ideas on this respect can be found in Droege & Braithwaite, 2015).

Nonetheless, can we conclude that those animals mentioned in the preceding section that passed the mirror test possess Edelman's primary consciousness (state of being mentally aware of things around)? Since animals, even those that do not pass the mirror test, perceive the world in relation to themselves, one can surmise they possess primary consciousness. And the fact that those passing the test are trying to modify something on their bodies suggests they also have higher consciousness of, at least, recognising their own acts. Now that we have endowed animals with some consciousness and self-awareness, what can we say of human babies?

Studies have been done on preterm and newborn infants, even on foetuses while in the womb. For instance, reactions of the foetus to the maternal touch and the sound of her voice have been reported, in that foetuses displayed more arm, head and mouth movements when the mother touched her abdomen and decreased their arm and head movements after mother said something. This indicates we already have some responses to the environment at a very early age, in fact before being born. Foetuses also display facial expressions, eye opening and yawning. Naturally, the foetus has a less developed brain than that of a newborn (Walsh et al., 2014). The foetal brain is smaller in size and smoother, yet the foetus displays movements and reactions. Interestingly, as an aside, let's mention that human development in utero is an analog of what is thought occurred during animal evolution: we begin as a single cell which multiplies, divides and becomes a complex multi-celled animal, all this occurring in an aquatic environment until we transition to "land" and breathe air.

Going back to the matter of the origin of cognition in the baby, when exactly in the brain of the foetus there is sufficient development for reaction, is unknown, but it is reasonable to postulate that a minimal number of interconnected cells is necessary even for reflex movement (Chap. 6 deals with the topic of how to go from simple reflexes to higher cognition). But can the observed foetal movements be programmed behaviours, perhaps of subcortical origin? (subcortical origin in this context means that the cerebral cortex does not cause the movements/actions as it usually does after birth; rather, some structures beneath the cortex are causing them, because the cortex is not fully developed in utero). It is in fact likely these are programmed actions. We only have to consider that many of our behaviours are programmed in the brain cell circuits. Our protective reflexes, like withdrawing the hand upon feeling fire, are an example of programmed actions in neural circuits. Not only reflexes but also our bodies are "mapped" in the brain. There is a representation of our body in the neural circuits of the brain, and this is why after losing a limb, people experience phantom limb pain; and even most remarkably, some children born without a limb or suffering phocomelia (malformations of the arms and legs) still feel the phantom pain, as if their brain knew what parts of the body should be present: the brain pattern corresponding to the limb exists even when the limb does not. This phantom limb pain is caused by remaining nerve pathways from the amputated or congenitally missing limb that continue sending signals to the brain areas (sensory cortices) that are supposed to take care, to feel as it were, that limb which is now absent, and these signals are interpreted as pain. Hence, it appears that we have a genetically determined body map in our brains, but one that can be modified by sensory experience or by accidents. The typical scenario occurs when, after limb amputation, there is a remapping in the brain sensory cortex area for that limb. Experts in the field of neuroprosthetics use this knowledge to create feedback systems with artificial limbs that are connected to the amputee's sensory nerves with electrodes. The new generations of prosthetics have the capacity for function and feeling, and it is a field that is advancing fast.

All this is mentioned so that we understand that the body is to a great extent mapped onto the brain, and thus, many sensations and behaviours are, in fact, sculpted on those brain circuits. If we combine this evidence with the incessant neural activity in the recurrent neuronal networks mentioned in Sect. 1.4, then it should not be too surprising that many behaviours and other psychological dispositions are just the manifestation of the activity in those brain cell circuitries. The brain is not initially a completely blank slate that will be filled up gradually with acquired experience, but a self-organised system (see Sect. 9.3 for what self-organisation entails) containing a vast repertoire of pre-formed neuronal activity patterns. Thus, our brains develop from conception with software and subroutines. Experience is mainly a process of matching these pre-existing neuronal dynamics to events in the world. Do you now perhaps feel a bit different about the notions you had on the concept of free will? Let us not jump ahead, because you will have to wait until Sect. 2.2 to release yourself from yet another illusion made by the brain, the presumed free will.

1.8.1 The Body in the Brain

A fundamental aspect of self-processing is the perception of one's own body. Corporeal awareness relies upon a large brain network including somatosensory, parietal (the coordinating hub) and insular cortices (all these very important brain areas of the neocortex). Some readers have probably seen images of the cortical homunculus (Latin for 'little man'), that typical cartoon that depicts the distorted representation of the human body on the cerebral cortex, based on the neurological map of the areas and proportions of the human brain dedicated to processing motor or sensory functions for different parts of the body. Those interested in this topic can read, among many other papers, Robert Melzack's, 2001 review on the concept of the neuromatrix, which, according to his proposal, "comprises a large network of neurons that generates patterns, processes information [...] and ultimately produces the pattern that is felt as a whole body possessing a sense of self" (Melzack, 2001). As we see, all these themes are intimately related to the perception of the self, and this is why the next few words explain very basic aspects of how the brain senses, or perceives, the body.

We need to feel our bodies to have a proper selfhood. Fortunately for us, most of the interoception (the sense of the internal state of the body) is unconscious, for we would go nuts if we were to feel everything going on in our body. We find first the brainstem, a very old (in evolutionary terms) region of the nervous system. While it may not be involved in creating art of musing about philosophical problems (these are tasks for the prefrontal cortex), it has the critical mission of orchestrating breathing and heartbeat, as well as monitoring the state of the organism. It checks the contents of the blood and cerebrospinal fluid, receiving signals from a variety of regions like the vestibular system for the detection of motion and the orientation of the body in space. Incidentally, this vestibular system is a key component of you being sick on a ship—a little advice on how to avoid seasickness: it occurs because when you are in your cabin, the information from your eyes tells you that you are immobile, but your vestibular system signals the opposite, which is true, that you are in motion, so if you go out on the deck and watch the horizon following the swaying motion of the ship, then both visual and vestibular information will agree, and your sickness will diminish. Without the vestibular system, you would have problems locating yourself in space and maintaining balance; thus, this small, tiny system that fits in the inner ear and its connections to the brainstem help your bodily self-consciousness. The brainstem receives signals as well from the spinal cord for proprioceptive signals (awareness of the position and movement of the body) and sensing the musculoskeletal frame, from the area postrema for monitoring chemicals in blood, from the periaqueductal grey matter to perceive pain and body temperature, from the nucleus of the tractus solitarius for an inspection of the states of the viscera in the body and the sense of taste, and from a few more neural areas, but these mentioned here are the main ones. In addition, the brainstem is responsible for the global arousal of the brain due to the projections of its neurons throughout the brain, because one of the main neurotransmitters (recall from Sect. 1.4 these are the chemical compounds that mediate neuronal communication) in the brainstem is acetylcholine—this transmission is termed cholinergic. Thanks to it, you will not fall asleep during the day when it is not needed (at night, though, cholinergic transmission slows down, and you, that is your brain, fall asleep). In conclusion, the state of the organism is continuously portrayed in many dimensions in the very ancient brainstem, and furthermore, it controls to a large extent the sleep-wake cycle, so a good brainstem will make you healthy and keep you alive.

As animals evolved and increased their body complexity, new nerves were needed to supply new brain areas beyond the brainstem for the sensing of their complex bodies. Hence, phylogenetically (related to evolutionary development) new pathways in primates exist that provide a direct thalamocortical input reflecting the physiological condition of the body, and in our brains, we have in addition to the brainstem several other areas like the insula, the thalamus and the somatosensory cortex, that serve interoception. To make a long story short, representations in the interoceptive cortex lead to a meta-representation of the state of the body in a very special brain region, the insula (more specifically the right anterior insula), that is associated with the subjective awareness of the 'feeling self'. For more details on this topic, one can read A. D. Craig's paper "Interoception: the sense of the physiological condition of the body" (Craig, 2003).

Now, perhaps a little apology here for those who may find these technical details somewhat tedious. While this is not a specialised book and is aimed at a general audience, many details are mentioned here and in other sections only to illustrate the complexity of the emergence of selfhood. It will lead to a better understanding of how we construct that perception we call personal identity, because it is not only that we reflect upon events and we conclude, like Descartes did a few centuries ago, that "I think therefore I am". This is a very superficial view on what the sense of self entails, and it is hoped that after these initial pages of the book, it is already apparent that the creation of our personal identities depends on the brain, body and environment that are intertwined in many fashions.

1.8.2 Our Starting Consciousness in Our Early Times

Going back to babies, preterm and newborn infants, whereas they are mostly asleep, display what can be called basic consciousness if only because they establish eye contact with the mother (although it is a minimal contact at this very early age), they display avoidance reactions to painful stimuli, habituation to a constant stimulus, and they show clear hearing and visual activity exploring the surroundings. Thus, the evidence suggests these early humans possess Edelman's primary consciousness, but not the higher-order consciousness in that they do not seem to recognise their own acts. This will occur soon as the social interactions—interpersonal transactions (self/other selves)—or "social cognition" develops in the mind of the child. One reason why these babies do not possess yet higher consciousness and a sense of self is given by neurophysiological evidence (obtained mostly using the so-called somatosensory evoked potentials) indicating that the thalamocortical connectivity—that is, the connections between the cerebral cortex and the thalamus, the latter being a major relay centre located more or less in the middle of the brain and through which sensory nerves transmit signals from the spinal cord and brainstem on the way to the cerebral cortex—is not fully established;. Therefore, there are some incomplete neurocircuits in these very young creatures. In fact, there are some areas of the brain like the prefrontal cortex that will take years to fully develop; parents who have been perplexed at the apparent wildness of toddlers and teenagers will find this neuroscientific explanation comforting, namely, that the prefrontal cortex is a major brain association area that, among many things, controls impulses, regulates the behaviour, or, if you prefer and simplifying a lot, reflects about implementing the right behaviour in the proper time. So, when do babies start recognising themselves as agents that make things happen, thus reaching the next stage of self-awareness?

An influential book by the Swiss psychologist Jean Piaget, *La Naissance de l'intelligence chez l'enfant* (originally published in 1936 and translated to many languages), reveals most of the questions we can have about the origin of intelligence in children. The experiments he carried out with his three children (it can be said he had his laboratory at home!) are so well explained that if you are one of inquisitive nature, you could try to reproduce some of those experiments with your own children. Piaget himself tells us the main message of his investigations (on his own children): "It can be said that the living being *assimilates* to himself the whole universe, as the same time as he *accommodates* himself to it". The two words in italics are the key notions that shed light on the development of selfhood and associated higher consciousness: we have to feel part of the world (assimilating, integrating into it), and we have to adapt to the world (accommodating, adjusting to it).

In greater detail, for those who may be interested to know a bit more about Piaget's work, here are the six stages that he identifies during the development of the child's intelligence:

1st – Use of basic reflexes (starting from birth).
2nd – First acquired adaptations, or primary circular reactions in Piaget's words.
3rd – Secondary circular reactions: making interesting events last.
4th – Coordination of sensory schemata and application to new situations (appears around the age 8 ~ 9 months).
5th – Tertiary circular reactions: discovery of new means to make things happen through active experimentation (appears around the age of 1 year).
6th – Inventions of new means through mental combinations.

The first two stages are elementary sensorimotor adaptations and not described as "intelligent" because these lack intention, whereas the rest are already intentional sensorimotor adaptations, that is, created with an intention to make things happen. Reviewing Piaget's book, you will find examples revealing that very young babies lack a sense of selfhood, or at least, they do not seem to perceive themselves as separate from their environment, such as when the child does not know he has to stand up in order to release and grab a towel placed under his body (the experiments about dog self-awareness mentioned in Sect. 1.7.1 used a similar scheme). It will take a bit of time, but things will occur fast: selfhood will appear, and the ego will start to increase, fostered by parents, friends, teachers and later in life colleagues and other events that will make that individual believe there is something special about him or her. Soon, we will see (Chap. 10) what really is special about us; in fact, we already had a hint in the closing comments of the previous Sect. 1.7.1 about how the intentions of our intellect may diverge from the genes' "purposes".

Using different terminology, but in reality sort of similar underlying concepts as the previously mentioned Edelman's primary and higher consciousness (it so happens that many times, scientists develop different theories about a phenomenon using specific and distinct words but that, once the essence is distilled, the theories say basically the same thing about that phenomenon), we can talk about what the neuroscientist Antonio R. Damasio proposes—that newborns and animals possess a proto-self, which would be the most basic level of self-awareness needing only rudimentary features like attention to objects or people around and detection of their significance. In Damasio's words, this proto-self is "a coherent collection of neural patterns which map, moment by moment, the state of the physical structure of the organism in its many dimensions" (Damasio, 1999). Hence, it is the brain activity patterns that are representative of the body's internal state. As the child's mind develops, we can then talk about core consciousness and the associated core-self, when the child or an organism becomes consciously aware of feelings associated with changes occurring in its internal bodily state and is able to recognise its own thoughts. The complete self, the traditional concept of self, Damasio calls the autobiographical self, the persistent collection of unique facts and ways which characterise a person, which echoes the preliminary definition of the self previously presented, the perception of the unity and continuity in behaviour and cognition.

Now that we have explored the origins and emergence of selfhood, let us proceed to understand how this perception of our personal identity further progresses thanks to the modelling done by our brains—models of our psychological traits and the surrounding environs—to inspect whether there are one or more selves that may be living in our heads and who of these, if any, may be in charge; and along the way, we shall demystify a few concepts.

Chapter 2
The Origins and the Fallacy of a Central Commander in the Brain: The Emergence of Agency and the Demise of the Concept of Free Will

Contents

If the previous section revealed that children at a very early age do not see themselves as agents and do not seem to possess an awareness of their own cognitive processes (self-reflection), this section will uncover the importance of experiencing ourselves in action, of developing a sense of agency, in the formation of our personal identities—agency is the feeling of being the agent of one's thoughts and actions. Furthermore, we will begin to see how the brain starts modelling our behaviour, how it starts to make a model of the self that contains the notion of a central commander, a brain region or mental construct controlling the behaviour. Is there really a brain region in control? And if so, how free is this central commander to act? How free is that will we are supposed to enjoy, and is it free will or "free won't"? Free won't is a term devised by neuroscientists who think that while some impulses to act are dictated by the subconscious, the conscious mind would still have the capacity to suppress or veto this urge to act, hence the freedom to stop behaving, or "free will-not (do it)". For some time, the topic of free will was taken to belong to the metaphysical realm, but it does not need to be that way, and in current times, there is plenty of neuroscience research on the theme—some thought-provoking experiments will be described below.

To start understanding these phenomena related to the sense of agency, it helps to recall in Sect. 1.8 the comments on the programmed neural circuits within the

J. L. Perez Velazquez, V. Nenadovic, *Being and becoming*,
https://doi.org/10.1007/978-3-030-78264-1_2

brain in conjunction with the fact that the brain never rests. Even when we are in deep sleep, there is substantial cellular activity recorded very nicely in electroencephalograms as rhythmic slow waves. This is the reason why non-dreaming sleep is normally called slow-wave sleep as opposed to rapid eye movement (REM) sleep episodes that are normally associated with dreams and with other types of brainwaves. Therefore, that continuous neural activity connecting brain regions results in many of our behaviours having a reflex-like character where not much cogitation is needed. Yet somewhere in our minds, we interpret those behaviours as caused by our free will, by our unbiased, unprejudiced determination. Is this perhaps another perception that should be questioned, like those of Figs. 1.2 and 1.3? Let us again emphasise that to a large extent, mental constructs are abstractions derived from the perception of patterns, and perceiving patterns in behaviour and cognition gives rise to the self and, with that, the sense of agency and the feeling of free will. It is now time for another step in the liberating path, releasing yourself from yet another illusion, that notion of free will. But before this, let us inspect what is needed to feel ourselves as agents, actors that make things happen around us.

The feeling of being "in control" of our actions is ubiquitous among people. This feeling results from a sense of ownership of our own body parts, from self-monitoring our actions or from a sense of agency, and along with these, a feeling of prediction of the movement outcomes is also required, which is achieved by on-line adjustment of movements that happens with or without awareness (in fact, most of these adjustments of our movements are unconsciously done). It is revealing to see how these feelings are created by the brain, and nothing better than to inspect some very interesting neuropathologies. Thus, the sense of ownership is disrupted with damage to the parietal cortex or in body illusions that do not require any pathology (we will see below in Sect. 2.1 one illusion which readers can experience at home). The sense of agency is abnormal in some deviations such as schizophrenia, where patients have hallucinations. Interested readers may want to consult C. Frith's paper "The self in action: lessons from delusions of control" (Frith, 2005).

Fundamental to the concept of agency is how we perceive motion. There have been many cognitive and neurophysiological experiments done that have studied how brains perceive motion—specifically of biological entities as opposed to scrambled, meaningless motion—and actions produced by others and those caused by ourselves. Motion perception is so important that there exist in the brain the so-called mirror neurons. It so happens that when you look at someone making a movement, say waving the hand, there are areas of your brain—the mirror system—that become active as though they were about to perform that waving. Specifically, neurons in the primary motor cortex are active after watching movements performed by others, a neuronal activity which could be unexpected because after all, you are not planning on waving your hand—it is then, when you really intend to do it, that your motor cortex becomes active and executes the action sending the cortical neuronal activity down the spinal cord nerves to the muscles that will move the hands. But the evidence demonstrates that some neurons in the motor cortex become equally activated during observation as well as during execution of motor tasks. Imaging studies also indicate other brain regions like the superior temporal sulcus, inferior parietal lobe and inferior frontal gyrus may be part of the mirror system in the human brain.

Now, you may think, why don't I wave my hand if my "mirror" neurons in the motor areas that control hand muscles and serve that purpose of motor activity are active watching the other doing it? There is in fact a condition, echopraxia, where the control of the routes to action has been lost, and as a result, these individuals involuntarily repeat or imitate another person's actions, and similarly, echolalia is the involuntary repetition of sounds made by others (these not only being core features of Tourette syndrome but also occurring in autism and schizophrenia). The neuroanatomical causes for these syndromes are under investigation. And in a non-pathological, everyday situation, this mirror system may be the reason why, as many readers have probably experienced, there is a tendency to imitate the same words or gestures made by people we watch either on TV screens or in front of us. While this mirror business has generated a great deal of interest in neuroscience and abundant research is addressing the implications of the presence of such mirror neurons in our brains, it is always instructive to look back at the evolution of animal behaviour and the actions of other animals around us and discover that, in the final analysis, our behavioural traits lie in a continuum with those of other beings. Hence, when we find that mimicry is a cornerstone of social learning in many animal species, this makes the finding of mirror neurons in our brains and our tendency to mimic gestures not that surprising. By the way, the mirror neurons may not only be related to perceiving and moving, because there are some cells in the human cingulate cortex—another cortical area involved in learning, memory and emotion processing—responding to watching and receiving pain. One wonders whether this mirror system may be responsible, at least in part, for that disposition called empathy.

Aside from these pathological conditions, one reason why we do not perform others' actions in spite of having some of our motor cortex activated seems to be the number of cells active: activation of a motor cortical area called M1 during imagined movements showed about 30% of the activation that occurs during effective movement. It could then be a question of mass activity: the more neurons active, the more likely it is their signals will travel down the spinal cord to move the muscles. But numbers may not be the only reason; there is also the matter with the neuronal coordinated activity. Some neuroimaging studies revealed that the spatial pattern of neuronal activity for a particular observed movement was different from that for the same movement when executed, suggesting that observed and executed movements are represented by either distinctly different subpopulations of neurons or that their activity is organised differently—as mentioned a few times in this book, the coordination dynamics of brain circuits, the organised activity among neuronal circuits, is a most fundamental aspect for appropriate behaviour. In fact, echopraxia has been observed in individuals with frontal lobe damage, which would lead to abnormal patterns of neural activity in the brain—remember that, as mentioned in previous sections, the frontal cortex and especially the prefrontal section is important for the "control", or regulation, of behaviours, for planning actions, judgment, impulse control and social behaviour.

Perhaps these considerations on the mirror system and on related aspects like echopraxia add one more nail to the free will coffin, for we start to comprehend how subtle and intricate and how many neurophysiological facets are involved in the presumed control of our behaviour. And to add one more nail to that coffin, "utilisation behaviour" is another illustration, although not exactly related to the mirror system, of the difficulty resisting what some brain areas command the muscles to do; individuals with this neurobehavioural disorder have no choice (more on this "choice" business later, in the section on free will) but to grab objects in view and start the appropriate behaviour associated with that object, but at an inappropriate time. For instance, if they see eyeglasses in front of them, they will grab them and put them on whether needed or not. Utilisation behaviour patients have difficulty resisting the impulse to operate or manipulate objects which are within reach in their visual field, and it has been linked to lesions in the frontal cortex—yes, once again, alterations in this fundamental brain area appear as a culprit for misbehaving. Hopefully, these few paragraphs have served to realise the intertwined character of perception and action and what is in between those two, which is usually called cognition, as if what lies in between had a different nature (this is another illustration of the enthusiasm for dichotomising and separating phenomena that can be hardly separated, and on this topic, we endorse the view of scientists like Ulric Neisser considering cognition as "all processes by which the sensory input is transformed, reduced, elaborated, stored, recovered and used", so basically sensing and acting are as many cognitive events as philosophising or solving equations). This entanglement between action and perception also teaches us that the distinction we normally make of sensation and movement as stimulus and response respectively is somewhat artificial and not a distinction which can be regarded as descriptive of anything, a notion already maintained by some scholars a long time ago, like the psychologist and philosopher John Dewey at the end of the nineteenth century who in 1896 published "The Reflex Arc Concept in Psychology" (in the scientific journal Psychological Review).

Whereas the perception of the body is fundamental for a sense of selfhood, it is far from perfect; we shall explore now how easy it is to fool ourselves about what belongs to our body.

2.1 Incorrect Body Perceptions: Illusions of Ownership and Out-of-Body Experiences

It is not difficult to mislead the brain in terms of body awareness. Let us start with one experiment that can be done at home, a somatic sensory illusion called the rubber hand illusion. One needs a "rubber" hand, or any fake hand made of any other material—but rubber makes a very good approximation to a real hand (the more it looks like a genuine hand, the better)—a small brush (a paintbrush works well) and something like a piece of cardboard or screen to prevent the participant to see his/

her own hand resting on the table, as this is done with the volunteer sitting at a table. In Fig. 2.1 and in this short video[1], you can see how it is done and its amusing consequences. The important matter is that the real hand has to be hidden behind that screen while the fake hand is placed in the position and space where the real hand would naturally rest on the table, and the volunteer can look at it. The person running the show will simultaneously and as equally as possible stroke the real hand and the fake hand with the brush (same frequency, same stroking motions), while the volunteer watches the artificial hand being stroked, and in a few seconds—in any case no more than 2 or 3 minutes—about two-thirds of the participants will feel like the rubber hand is their actual hand (the illusion does not occur in all people; I did experience it once, which lasted only a few seconds and was a, let us say, very intriguing experience); in other words, they will disembody their hidden, real hand and adopt the rubber hand, as the brain links the sensation of the real hand being stroked with the observation of the fake one. At the moment the stroking ceases or it becomes asynchronous between the fake and real hand, the illusion vanishes.

This simple experiment reveals that what we see and feel determines our perception of ownership of body parts. The brain combines information from the senses to create a feeling of body ownership, not only in this amusing experiment but all the time in real life; in the case of the rubber hand experiment, the brain combines the visual information with the touch sensations to the point that we feel that the rubber hand must be part of our body.

And at this point, we can start talking about a procedure that, as we will see in future pages, provides information about how we make up features associated with our selves: electrical stimulation (or by other means, stay tuned) of the brain. As such, this deceptive feeling of ownership in the rubber hand illusion is either lessened or strengthened depending on what brain area is stimulated: the former happens if repetitive transcranial magnetic stimulation (rTMS) is applied over the parietal cortex and the latter if it is the extrastriate body area that is subject to rTMS. For those interested, we will note that the extrastriate body area is involved

Fig. 2.1 The rubber hand illusion. Participants will experience a shift in perception, feeling that a fake hand placed on the table before them is their own; it will occur in about 70% of individuals. (Image in the public domain)

[1] www.youtube.com/watch?v=nzF_DfOafKw

in the visual perception of human body—there is another cortical area, the fusiform face area, that is implicated mainly in the perception of human faces—and the parietal lobe which we have met several times before is a multimodal association cortical area which assists in the integration of body image and coordinates motor output. Hence, either damage or alterations of activity (for instance by rTMS) of the parietal area produces inability to properly combine sensory input and motor output and may induce some interesting incidents that we will see below like impaired volitional movement execution, and TMS is a non-invasive form of brain stimulation in which electromagnetic induction is used to cause electric currents at specific brain areas, therefore changing the neural activity in those areas, and besides being used as an experimental tool, this method has therapeutic uses in the treatment of neurological syndromes. Going beyond the hand, with today's virtual reality technology, there are more possibilities, like inducing illusory ownership of whole virtual bodies (Slater et al., 2009 shows how virtual limbs and bodies can come to feel like real limbs and bodies).

Physicians have taken advantage of how easy it is to fool the brain in this type of body illusions. There exists mirror therapy, widely used in rehabilitation therapy. This treatment was originally created to relieve phantom limb pain, explained in Sect. 1.8, and used as well in the treatment of other kinds of one-sided pain and loss of motor control, for example, in stroke patients suffering from hemiparesis (weakness or the inability to move on one side of the body), where it has been shown that it enhances motor recovery in post-stroke hemiparesis. To implement this therapy, the set-up consists of a mirror that is placed between the arms or legs of a person who has one of those limbs affected by some pathology (after stroke, injury etc.), so that the image of the moving healthy limb gives the illusion of normal movement in the affected limb. As we can start to suspect from the abovementioned illusions of ownership of body parts and this mirror therapy, it seems like our brain prioritises visual feedback over somatosensory or proprioceptive feedback; it apparently believes more what it sees than what it feels through sensory perception. The neurophysiological reasons for the success (it works in some cases, it does not work in all patients, but nothing is perfect) of the mirror therapy are still unclear, and perhaps it is no coincidence that scientists think the mirror neurons—those talked about a few paragraphs above—are involved. In general, it is thought that different brain regions responsible for movement, sensation and pain are stimulated by simply watching in the mirror the healthy limb moving normally, so possibly these mirror neurons become activated by the observation of movement in the mirror and help in the recovery of the affected limbs perhaps by stimulating the motor processes which would be involved in that movement with the paralysed or injured limb.

Always remember, the main take-home message of these sections, which will help understand the following sections on how the sense of self is created and altered, is that the brain is not a rigid network of neurons set in a given arrangement for the whole duration of your life, rather it is a plastic, flexible organ that changes frequently its activity and anatomy depending on what it perceives to adapt to the environment. Let us now inspect other remarkable body illusions, some of which may have an esoteric scent, yet the neurophysiological bases are relatively well understood.

2.1.1 *The Mind Out of the Body: Out-of-Body Experiences*

Out-of-body experiences (OBEs), generally called autoscopic phenomena, are visual hallucinations during which subjects have the impression of seeing a second own body. This phenomenon is related to near-death experiences (NDEs). Stories abound of patients who go into cardiac arrest (thus are about to die) and experience themselves floating around the room out of their bodies. When the patient regains consciousness, he/she describes the experience to an awed audience of doctors, nurses and relatives. But, do you know you could experience one of these OBEs without cardiac arrest, without being near death? We shall see how, and this will demonstrate that, as many times said in this book, (almost) everything is in your brain. But before spoiling, at least in part, the esoteric side of this theme, let us explore a bit this intriguing world of NDEs. Now, some readers may be wondering why this topic is treated here in a text about the self and consciousness. The reason is that these phenomena illustrate what we have been already discussing in previous sections: how brains model, or configure (to use a term of the digital age in which we live), our identities and our reality.

One can read numerous reports of patients who underwent surgery or who were about to die not necessarily during surgery, and reported, after "coming back", that they had an OBE, or another set of experiences common in NDEs such as seeing bright lights or light at the end of a tunnel, a feeling of ineffability, or peace and well-being. Of note, NDEs have been known for a long time, for example, sixth-century Pope Gregory the Great collected examples of NDEs in his book "Dialogues". These are profound psychological events with transcendental and mystical elements occurring to some individuals close to death. One famous case is that of the singer and songwriter Pam Reynolds who, in 1991, had surgery to deal with a large brain aneurysm, which had to have her body temperature lowered, her blood circulation dramatically decreased and her head drained of blood. The operation, which lasted about seven hours, was successful, and Reynolds recovered completely. Later, she described hearing a musical note and then "popping out of her head". She recalled watching the doctors perform the operation, and she described some details of the procedure that matched those reported by the surgical team. Her case was studied, and in the opinion of the anaesthesiologists, this was a case of anaesthesia being too light such that she had partial awareness of things around her. Other experts, though, have questioned this light anaesthesia explanation, because that surgery required very strong anaesthesia. Nevertheless, in spite of being severely sedated, she was still able to perceive, of course very vaguely, some sounds and movements around her and, upon waking from the anaesthesia, associate and elaborate upon the sensations she had experienced during the operation with her existing knowledge and expectations (and here, please remember Sect. 1.1 and Figs. 1.2 and 1.3 showing that what we experience is in part due to what we have in mind and what we expect). The question remains, as advocates of the afterlife would maintain, that if her brain was unresponsive, then how come she (her brain) could be aware of those things during the surgery to be elaborated on and explained later;

in fact, during NDEs, thinking seems to be lucid but occurs during periods of (apparently, see below) absent cerebral function.

We have no room in this book to go deep into the, many times controversial, issue of death (in clinical settings), and, in case you did not know, the notion includes not one but several "deaths", and all this being the product of our technological advancement that has created an artificial dissociation of vital functions; for instance, today, there are mechanical ventilators and extracorporeal membrane oxygenators that allow continuation of ventilation and circulation even in brain death. Just for your information and perhaps amusement, and to see that even dying is not a trivial task in our very bewildering and artificial world, here you have the types of death talked about by healthcare practitioners:

- Whole-brain death (the irreversible cessation of the critical functions of the entire brain)
- Brainstem death (the loss of consciousness combined with the loss of capacity to breath, recall that the brainstem was described in Sect. 1.8.1, and it is a fundamental part of the nervous system that maintains us alive by controlling breathing, blood pressure, heart rate and in general monitoring the state of the organism)
- Then there is the "higher-brain formulation" of death (loss of "higher" cognitive functions served by the neocortex, and naturally, this definition of death would apply only to Homo sapiens)
- And finally the circulatory formulation (the irreversible loss of circulatory activity, but some argue this is a criterion sufficient but not necessary for death)

One thing behind all this death-mess that may help explain NDEs like that of Reynolds is that even if there was a flat line in brain recordings during the operation, this would indicate loss of neural activity but not a total loss, because we have to consider that our methods to record brain activity from the scalp are somewhat crude and not sensitive enough to detect minute neural activity that may be going on in an otherwise quasi-dead brain. For starters, scalp EEG records mostly cortical neural activity, but not that activity that maybe present in deep brain structures. Besides, a sufficient synchronisation of neuronal activity is needed to record activity from the scalp, so if the neurons are hypoactive there may not be enough synchrony to be recorded at the scalp, although that hypoactivity may be enough to sustain awareness. Let us now leave this digression on the complications of dying in our time and age, and go back to see that the OBEs do not need to be necessarily involved in near-death occurrences.

Because as it happens, experiences indistinguishable from NDEs can be elicited by electrical stimulation or drugs, a fact which may be disappointing to some inclined towards more mystical perspectives. Thus, OBEs were evoked by the pioneering intracerebral stimulations of neurosurgeon Wilder G. Penfield in the 1940s. As it is customary previous to some brain surgery, Penfield's patients had to undergo some sessions where parts of their brains were stimulated by current injection via electrodes while they were awake in order to know which brain regions should not be touched or removed by the surgeon. During the stimulation, the patients inform the surgeon what they feel, see, hear, etc. In these sessions, he mapped out more precisely the brain neocortex in terms of the function of its parts—recall that we have already talked about the sensory and motor cortex, visual and parietal cortex, etc. In addition to mapping the cortex, he found that some stimulations could trigger memories and perceptions in his patients, and interestingly, when an area called the temporo-parietal junction (TPJ) was stimulated, an OBE was sometimes induced. In the words of two of his patients, "I feel queer, as though I were floating away", and a second one said, "Oh God! I am leaving my body."

Following Penfield's findings, let us describe in a thought experiment how you can experience an out-of-body illusion (disclaimer: please don't attempt this at home!). You are comfortably sitting on a chair and allow a skilled neurosurgeon to introduce an electrode into that aforementioned area of your cerebral cortex called the temporo-parietal junction (to see the location in the brain, look at Fig. 2.3). Then, your surgeon friend starts sending an electrical current through that electrode, playing with different current intensities, until one of those stimulations activate the necessary neuronal networks for you to feel you are leaving your body. No need to be in cardiac arrest to experience this illusion.

Typical OBEs are characterised by disembodiment—that is, the self feels located outside one's body—and the impression of looking at the world from an elevated visuospatial perspective along with seeing one's body from this perspective (see Fig. 2.2). Therefore, we can see that this phenomenon challenges the model of the self that the brain (continually) makes in that the spatial unity of self and body is completely altered. But there are other forms of these autoscopic phenomena, namely, autoscopy that consists in seeing a double but without disembodiment (the person knows where his real body is but perceives another in front), and heautoscopy that is seeing a double and one's own body at the same time and having difficulty localising in what body the self is located. For clarity, these various forms of disturbed self-processing are pictorially shown in Fig. 2.2. The case of heautoscopy

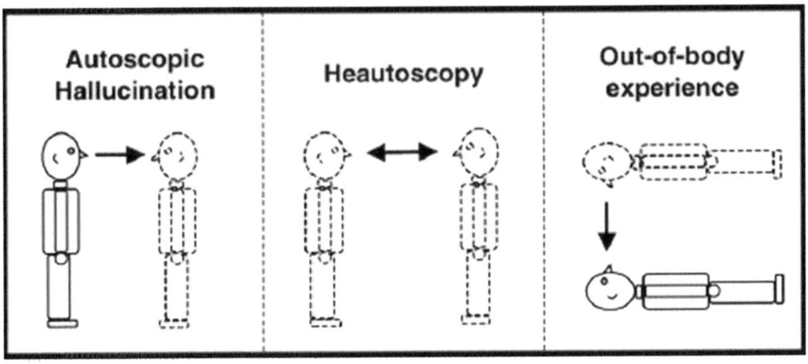

Fig. 2.2 The three autoscopic phenomena. The experienced physical body is indicated by solid lines and the experienced position of the disembodied (illusional) body in dashed lines. The arrow points away from the location from which the individual feels to be looking, so, for instance, in the autoscopic hallucination, the patient feels he is in his true own body but sees a double body in front, whereas in the OBE, the patient feels he is out of the real body and looks at it from an elevated perspective. (See text for more details. Reprinted from Blanke & Mohr, 2005)

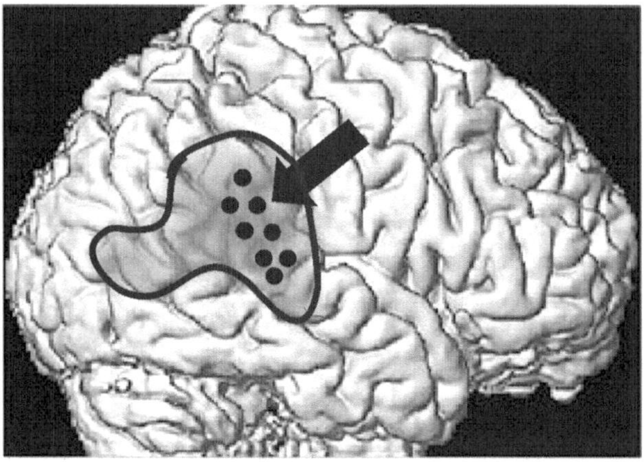

Fig. 2.3 The brain areas, in grey surrounded by the black line, around the temporo-parietal junction (TPJ) that elicited out-of-body experiences in a patient with implanted electrodes (black dots) in that region

is very disturbing for the patients because they cannot locate their selves (or themselves) in any of the two bodies they perceive (see figure legend; the illusory bodies are drawn with dashed lines).

As mentioned above, the temporo-parietal junction, or TPJ for short, seems to be involved in the out-of-body experience not only because its stimulation causes the illusion but also because patients with damage to the TPJ tend to experience the phenomenon (reviewed in Blanke & Arzy, 2005). What is peculiar about this cortical region? The TPJ, like its name indicates, is the region where the temporal and parietal cortex converge, and functionally, it incorporates information from the thalamus and the limbic system as well as from the visual, auditory, and somatosensory systems; therefore, it is in a perfect position to integrate multisensory information from the external environment and from within the body. For these reasons, the TPJ is a core region for self-processing and perspective taking. Those interested to know where this brain region lies can take a look at Fig. 2.3 that shows the location and as well the spots where electrodes implanted in that region evoked OBEs in a patient reported in the paper by De Ridder et al. in 2007. The patient, who had electrodes implanted in his right TPJ for suppression of intractable tinnitus, began to experience the illusion within 1 second after starting the stimulation, and it included a perception of disembodiment as if he were located about 50 cm behind his body (note this time it was not above the body as it is the more usual out-of-body experience). Other reports of patients whose TPJ was electrically stimulated indicate that the individuals felt or thought that they were either larger, smaller or outside their bodies.

But it is not only electrical stimulation of brain tissue that causes these illusions. Other insults like anoxia (lack of oxygen) or hypercapnia (abnormally high level of carbon dioxide in the blood) can produce typical NDE phenomena such as seeing brilliant lights, reliving past memories and having OBEs. This occurs because during anoxia, for instance, the brain visual cortex experiences a dysinhibition, that is, a higher than normal activity. This enhanced excitation of visual areas has been suggested as an interpretation of seeing bright lights or tunnel-like perception during NDEs. Interestingly, hypercapnia has a powerful effect on blood vessels in the brain, causing them to dilate which will bring increased blood flow to the brain, thus enhancing neuronal excitability.

Drugs can cause these phenomena too: ketamine, which is normally used as an anaesthetic but is also consumed recreationally, causes feelings of detachment from one's body and sensations of illusory movements. Other hallucinogens like LSD or psilocybin have been reported as well to induce these illusions. And even some mechanical induction has been known to develop into OBE-like states, such as the strong g-forces experienced in high-performance aircrafts, perhaps because that causes blood to drain from certain parts of the brain, hence causing temporary anoxia, and we saw above that, anoxia triggers OBEs. And why not, since (almost) everything is in our minds, mental induction of OBEs have been reported after meditation and during sensory deprivation or overload, perhaps because these latter procedures tend to cause confusion and disorientation.

Finally, let us mention that OBEs and mystical experiences appear in other contexts like in syndromes such as epilepsy: in the words of F. Dostoyevsky "I have really touched God… You all, healthy people, can't imagine the happiness that we epileptics feel during the second before our attack". He was describing the aura that can precede the epileptic seizure (auras can occur too before a migraine). As an aside, it may be interesting to some readers to know that preceding the epileptic seizure, or ictus, which is characterised by aberrant neuronal activity, there is already some abnormal neuronal discharges that cause patients to feel what in clinical parlance is termed an aura, which includes a variety of sensations like flickering lights, blurry vision, partial vision loss, feelings of deja vu, panic, etc. Whereas the Russian writer was talking about the moments preceding the seizure, interestingly, other people experience their whole seizure as a happy, exciting or even euphoric time. And there are other neurological conditions that can result in OBEs, like migraines, further strengthening evidence of physiological and not paranormal explanations. In short, there is abundant neurophysiological evidence of altered neuroelectrical activity as the roots of these phenomena. While it is true that mystical experiences can have a profound effect on the person and hold a great fascination in religion and popular culture, demystifying these experiences can be to some extent disappointing. However, finding a neurophysiological explanation does not change the lived experience.

The thought experiment previously described used an invasive stimulation (invasive means that an electrode, or any other utensil, penetrates the skull and brain tissue), but we saw that there are non-invasive methods for neurostimulation, like the transcranial magnetic stimulation (TMS) described above in Sect. 2.1; thus, an intriguing possibility is that if it were possible to stimulate the TPJ using TMS, then one could experience an OBE with no danger of surgeries or electrodes invading the skull. Isn't that interesting? Has anybody thought about it? Indeed, there is nothing new under the sun, so in fact, some neuroscientists proposed, already in 2007, that it may be feasible to induce OBEs in healthy volunteers through non-invasive TMS over the TPJ (Blanke & Thut, 2007). To our knowledge, this has not been done yet (or at least published), but what is known is that TMS applied over the motor cortex causes a sense of movement in the absence of a real movement. And if the experiment could be done and you wanted to volunteer, do not worry that you will not come back to your body, for you will never be out; this is just one of the many illusions that we have seen in this text. Always remember that the brain is the ultimate, supreme illusionist.

2.1.2 A Mini Science Project: Elucidating the Intriguing Phenomenon of OBEs in Blind Individuals

In closing these sections on NDEs and OBEs, it is worth mentioning, if only to motivate some controversy at the border of pure science and mysticism, one fascinating observation. It so happens that OBEs have been reported in blind individuals and particularly in those congenitally blind (from birth). The book by K. Ring and S. Cooper *Mindsight: Near-Death and Out-of-Body Experiences in the Blind* reports these intriguing cases. But how can this be? Are individuals blind from birth reporting detailed visual perceptions during OBEs or NDEs, in spite of not having any concept of what light or colours are? Not even understanding what seeing means, as one patient said during the interview admitting he could not explain how he had the perceptions he did because "I don't know what you mean by seeing". Before we give in to the natural tendency or even desire in many to believe in the miraculous, let us not leap too quickly to conclusions and let us play scientists. We invite you to become, if only for a couple of minutes, a scientist.

So put on a white lab coat and accompany us to see whether we can interpret these findings based on the information we have, what we already know about brains and sensory organs, some things we have already covered in previous chapters. The first task is to collect clear evidence, because many of these reports have been discovered to be complete fabrications (fake news proliferate in this type of esoteric or parapsychological matters, so one has to be careful). We scientists spend a good deal of our time searching for information, collecting good evidence that is normally found in publications in scientific journals or books or even in talks and conferences. While many of the news about the topic of OBEs in the blind are just occasional anecdotal reports, there are some serious publications that merit attention.

One is a paper by Kenneth Ring and Sharon Cooper (the authors mentioned two paragraphs above) in 1997, "Near-Death and Out-of-Body Experiences in the Blind: A Study of Apparent Eyeless Vision". They interviewed 31 subjects (blind and congenitally blind), who believed they had undergone either an NDE or an OBE not necessarily related to any near-death incident. Of major interest are those congenitally blind, who would not have had any visual input, because many of the interviewed subjects had had limited vision or were not blind from birth, and in these cases, their brains already accumulated abundant visual information to re-create images under different circumstances. In fact, it is known that most blind people who lose their sight later in life can dream visually, that is, they have visual imagery in dreams. Visual perception is one thing and *visual imagery* another; the latter is the faculty whereby we can re-visualise a visual item from memory as it occurs in dreams with visual content. What is the main evidence gathered from those 31 individuals in the Ring and Cooper's study? That most of these blind subjects report the same kinds of visual impressions as sighted persons do when they describe NDEs and OBEs, such as seeing a medical team at work on one's body etc. Hence, these people, some of whom have never seen anything in their lives and have no clear concept of what seeing means, seem to "see" during their experiences, as paradoxical as this may sound.

Now let us try to gather some things we know to see whether we can explain this remarkable phenomenon. We have seen in past chapters that the brain has a map of the body; recall from Sects. 1.8 and 1.8.1 that there are people born without a limb and yet who suffer phantom limb pain as though the brain was missing a limb that was expected to have been present at birth. Recall too the mirror system, those mirror neurons that become active when they "anticipate" they are going to be performing an action, yet the movements are never executed. All this indicates that, conceivably, the visual cortex, that part of the brain that "sees" or interprets the inputs from the retina in the eyes, may be prone to act, to perform its function of seeing, even though the neural networks may be pathologically altered or the nerves from the eyes may not be reaching it. If this is the case, if by any chance there were hyperactivity in those neuronal networks of the visual cortex, it would not be implausible to predict that those networks would "see" something that is not there, that is, the person would hallucinate. In this regard, remember as well that injuries to the tissue like anoxia (lack of oxygen) produce hallucinations and NDE-like visions because, as it is well-known, at the beginning of the anoxic episode, there is hyperactivity and disorganised activity of the neurons, which will be followed by periods of *hypoactivity*. We mention this detail about neural hyperactivity followed by hypoactivity in case some readers may be conversant with the fact that after stroke or ischaemic injury, which have associated anoxia, there is decreased function of the affected areas in the brain reflecting the neuronal hypoactivity, but again, at the start of the injury there is the opposite, greater neuronal activity that normally occurs in a disorganised manner; hence, the possibility to hallucinate arises. We have also mentioned in this book several times the fundamental importance of having organised neuronal activity for proper brain function, and recall in the preceding section the comment on epileptic auras which have associated strange sensations due to the abnormal disorganised neural activity before seizures. Good cognition needs appropriate neuronal coordinated activity.

There are more clues. It is also established that there is unconscious perception during sleep or in altered states of consciousness like the vegetative and minimally conscious state (for those who care, the term vegetative state is changing to unresponsive wakefulness syndrome). Did you ever wonder why the crying baby awakens her mother while another loud noise may not? During deep sleep (as well as during the pathological vegetative and minimally conscious states), the brain continues to receive sensory inputs from the outside, but the activations of neural networks normally do not extend beyond the sensory areas. So in these cases, the neural activity, or excitation, caused by sounds is restricted to the brain auditory cortex. This is different from the awake state, where excitation of the auditory cortex then excites other connected brain areas where this information is integrated with inputs from other sensory systems and activates memories. Thus, during wakefulness, the activity extends beyond the auditory cortex to association areas like the frontal cortices, and therefore, we become conscious of the sounds. Clearly, for the sleeping mother, the sound of her crying baby reaches the threshold to activate the brain beyond the local auditory area resulting in attention to the sound, and she wakes up; the evolutionary importance of a female attending to and protecting her offspring may explain a mother's ability to sleep through her partner's snoring, yet jolt awake at her infant's smallest cry.

What all this mentioned in the last two paragraphs means, in terms of our current problem in hand, is that there may be additional brain areas, not necessarily the visual cortex, that participate in the generation of the apparently visual hallucinations upon receiving input from retinal nerves. In fact, and to finalise collecting evidence that may help us explain the OBEs in the blind, it is known that axons in the optic tract (these are the nerves carrying visual signals from the eyes) terminate in four nuclei within the brain, some of which relay the information to the visual cortex, but other parts do not; as such, the lateral geniculate nucleus of the thalamus will transmit the visual signals to the visual areas, but then we find those axons from the retina also contacting the superior colliculus of the midbrain that controls eye movements, or the pretectum of the midbrain for the control of the pupillary light reflex, or the suprachiasmatic nucleus of the hypothalamus which controls diurnal rhythms and hormonal changes. So we see that several brain regions receive the eye signals, and whereas only the visual cortex can properly interpret those in order for us to "see" clear images, the other brain sections do perceive what the eyes see although may not be able to interpret or form the images in a proper fashion as the visual cortex does (recall the blindsight phenomenon from Sect. 1.1, in which patients suffering from extensive cortical blindness appear to be able to "see").

And finally, another piece of useful information for our task is the report that congenitally blind people have visual contents in their dreams, and some are even able to draw these images they claim that they visualise during dreaming (Bértolo, 2005). The issue of visual imagery during dreams in blind subjects has been controversial for some time. As it happens many times in science, for a time, it was concluded that there was no convincing evidence of visual perception in the blind, and while it is true that many of the dreams of the congenitally blind are without visual content—their dreams contain mostly sounds, touch sensations or emotional experiences—it has been reported recently, as aforementioned, that some do experience visual imagery during periods of dreaming which were accompanied by electroencephalographic features (recorded using EEG) indicating enhanced activity of their brain visual areas, that is, their visual cortex was becoming active at the time they were allegedly dreaming. Other studies provided more evidence for the fact that the congenitally blind are able to visualise, in spite of having never experienced sight, and the explanation of this phenomenon by one of these investigators parallels that which we mentioned above involving activity in brain visual areas: "subjects who have never had visual experiences can have dreams with virtual images that are probably mediated by the activation of the cortical areas responsible for visual representations." (Lopes da Silva, 2003).

According to these results reviewed in the last few paragraphs, a scenario that emerges is that enhanced neural activity in the brain visual areas, which may have been caused by episodes of anoxia or other injuries that cause neuronal hyperactivity and that are associated with the OBEs, is able to form some sort of virtual image that is interpreted by the individual as seeing something. If besides this perception or feeling we add the conversations the patient may have had after the episode with relatives or friends and the possible elaborations in the mind of the patient of the visualised images and experiences, then we could expect the patient to give us an account of the OBE very similar to the account of a sighted individual. Yet, some differences may be apparent, specifically that the description provided by the blind person should be a bit more ambiguous, fuzzy, if only because he/she is describing sights of things he/she never had seen before, plus remember that even the concept of seeing is foreign to the congenitally blind. Is this how the reports of those patients interviewed by Ring and Cooper sound? One conclusion in their paper is that "it seems more and more difficult to claim that the blind simply see what they report. Rather, it is beginning to appear it is more a matter of their knowing, through a still poorly understood mode of generalised awareness based on a variety of sensory impressions, especially tactile ones, what is happening around them." As an illustration of this, one patient told them that "she was never able to discriminate colours as such, but only 'different shades of rightness', about which impressions she could only wonder afterwards whether they represented what sighted people meant by colour". Hence, we can also notice there was some elaboration of her experience after talking to others. The authors also conclude that, talking about one of the patients named Sarah, "The story of Sarah implied that she really could see during her NDE in the way that a sighted person might. We have shown this is an unwarranted inference. What seemed like an analog to physical sight really was not when examined closely". Furthermore, the authors also acknowledge that there may have been retrospective reconstruction, elaborations of the stories based on what others told the patients which helped them reconstruct a plausible account based on all their prior expectations, experiences and knowledge—familiarity with hospital routines, overheard conversations or other sensory cues at the time of surgery—and explanations of others; in the words of the authors, "by the time we encounter them [*the patients*] the stories have long come to be expressed in a particular linguistic form, and that form is a language of vision, since our ordinary language is rooted in the experiences of sighted persons and is therefore biased in favour of visual imagery". Indeed, for visual animals like us, reporting any of our experiences normally use this visual language.

In closing, it seems that the explanation for this phenomenon involves a complex multisensory awareness that generated the visual imagery that blind individuals reported they had during OBEs. Conceivably, then, the phenomenon occurred when some non-retinal-based mechanisms that combine modalities like tactile qualities are added to some elaboration after discussing their experience with others, which resulted in accounts of apparent visual experiences during the OBEs. You can take off your lab coat now, as our investigation has ended. Have we found the complete explanation for this phenomenon? Perhaps, but most likely not, for reality is most times stranger than we can fathom. As we can read in the Ring and Cooper's paper: "What the blind experience is more astonishing than the claim that they have seen." Nevertheless, as an intellectual exercise, it was worth doing. Now you see how we scientists spend all our time.

2.2 Free Will, or What Is Free in that Will?

So you still think you have free will? The previous sections have covered how we start developing a sense of agency via the brain sensing the body and its movements, agency which is fundamental in the making of the selves and the expression of the will, free or forced. The feeling of free will, of our ability to perform unbiased decisions, is extremely resilient and widespread among people. It was none other than the great psychologist and philosopher William James who said, in a diary entry in 1870, "The first act of free will is to believe in free will". We have already started to put some nails in the coffin for this notion of free will at the start of Chap. 2 when discussing echopraxia and utilisation behaviour. Now we can continue adding nails by going back to that thought experiment of Sect. 2.1.1. This time, allow your neurosurgeon to stick the stimulating intracerebral electrode in another brain area, in the right parietal cortex. Sending some low-intensity current through that electrode, the neurosurgeon will cause you to claim that you have a desire to move the contralateral part (opposite side) of your body but without actual action performance, and upon increasing the stimulation intensity, you will finally claim that you have moved what you so much "wanted" to move. In reality, as your surgeon friend will tell you, you never moved anything; it is all an illusion produced by that stimulation. If the intracerebral electrode had been placed in the left parietal area, you would have claimed an intention to move your lips and start talking. In Fig. 2.4, you can see the brain cortical areas where the stimulation is performed and your claims—what you feel or think is happening—together with the responses from the surgeon-scientist. Now he blocks the view of your arms and legs, and he inserts the electrode into your premotor frontal cortex and starts again to send current to stimulate the neurons in that area, and as a result, you will claim that nothing happens, that you do not wish to move anything and nothing has moved. But when you are offered visual feedback by allowing you to see the rest of your body, you will watch your limbs moving during his stimulations: the activation of those neural networks in the premotor cortex is evoking actions of which you are unaware, and you may even continue denying performing the movements.

This thought experiment is based on the observations reported in some patients with intracerebral electrodes who were undergoing awake brain surgery. The main results of these studies revealed specific cortical areas that process information related to the intention to perform movements and the awareness of the motor

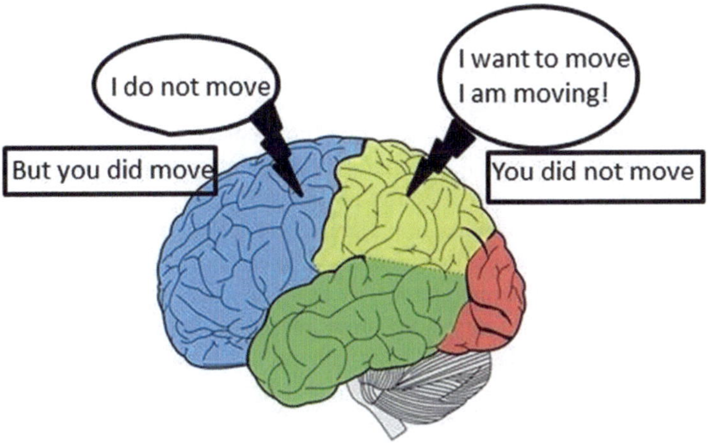

Fig. 2.4 Manipulating your will. Shown is what you would say (in circles) if your frontal cortex (blue) or parietal cortex (yellow) were stimulated with current (black arrows) from an intracerebral electrode located in those regions. In the rectangles is shown what the researcher stimulating your brain will say, representing the true events. The other two cortical areas represented in colour are the temporal lobe (green) and the occipital cortex (red, this one also called the visual cortex as it is devoted to the processing of the visual inputs from the eyes). Under the cortex, there is a multitude of deep brain regions that are connected to these cortical lobes, some fundamental for goal-directed actions that we will see in the text, like the basal ganglia introduced in the following section

actions (the experiments are reviewed in Desmurget & Sirigu, 2009) and shed light on the coordination of neural activity between cortical areas necessary for the experience of intentional movements. If you want to read technical reviews on the brain networks and the neurophysiology involved in voluntary actions, see Haggard, 2008, and Hallett, 2007.

Free will as it is commonly understood, the capacity to make decisions in a completely conscious and unbiased manner, does not exist. After what we have already covered in this text, hopefully, it has become clear that it is basically impossible to make unbiased decisions. What exists, instead, is our capacity to choose, but those choices are, I am afraid, determined by what you already have in your mind and what you perceive at the moment of making the choice—this is the essence of J. D. Schall's words "If we ask whether we are free, the kind of answer we want may not be possible. A better question to ask is: do we make choices? The answer is certainly yes. Are our choices constrained? Yes" (Schall, 2001).

One brain region will force neurons in another region to have a certain activity that in turn will change the activity in other connected neural networks resulting in the end in you making the choice, deceptively free only because you have no conscious access to most of those computations among neural networks. Thus, the illusion of consciousness will appear due to the belief that we are intrinsically informed about how our minds/brains cause actions by the observation of causation and by following our chain of reasoning. But again, that chain involves many steps that are beyond our awareness, as the philosopher Maurice Merleau-Ponty told us "What misleads us on this is that we look for freedom in the voluntary deliberation which examines one motive after another and seems to opt for the most convincing" (Merleau-Ponty, 1945).

The brain stimulation studies in patients during awake neurosurgery described above demonstrated that the experience of volition can be felt erroneously. As illustrated in the thought experiment, parietal lobe stimulation in these patients induced a will to move but without actual action performance, and increasing the intensity of the stimulating current resulted in patients claiming they had produced a movement that in reality was never done. This (false) conviction of having moved was probably the effect of activating connected areas in addition to the parietal regions due to higher intensity of stimulation: more neuronal chains were involved in the activations, more brain areas may have become recruited, and some may have reached neural networks responsible for that feeling of agency talked about in these sections. Thus, more widespread neuronal activation may lead to, erroneous in these cases but correct in normal life, perception in movement awareness/monitoring (sense of volition and movements). Particularly in the neurostimulation experiments described, the current sent through the electrode may have resulted in enhanced coordination, perhaps synchronisation, of activity among those local neuronal areas that were stimulated and the coupled regions, albeit artificially enhanced using electric current—but, as there has not been any assessment of synchrony in neuronal activity in these studies, this remains a speculation. Note that while in these experiments the trigger for those neural chains activating one another was artificial via the intracerebral electrode, in real life, in our everyday actions, there could be identical neural chains triggered not by inserted electrodes but by the intrinsic cellular activity within the many brain circuits that inform us when we make an action or, in other words, tell us we are agents performing actions. The sense of agency is thus born at almost every step of our lives, and it matters little whether it is due to artificial neurostimulation or to our own endogenous neural activity.

On the other hand, frontal cortex stimulations (the premotor cortical areas shown in Fig. 2.4) evoked actions in the individuals, but these were unconscious. The patients denied that they had performed the movements, unless offered visual feedback of their limbs performing the movements. This is a fascinating, albeit slightly frightening puppeteering example (Fig. 2.5 will introduce you to some puppeteers in your brain). Hence, motor awareness seems to emerge from predictions made about the movements rather than from the sensory feedback caused by the movement itself (called proprioception, the sense of self-movement and body position). The two cortical regions where electrodes were placed in these studies, the frontal and parietal cortices, are two fundamental brain regions implicated in executive functions. The frontal areas produce the signals that are sent to the spinal cord to produce the movements, and the parietal lobe (one important coordination hub) is a multimodal association cortical area which assists in the integration of body image and coordinates motor output.

The general theme and take-home message that emerges from these and some other neurostimulation studies (which we did not cover for the sake of brevity) is that of certain coordinated activity in a distributed set of brain areas which gives rise

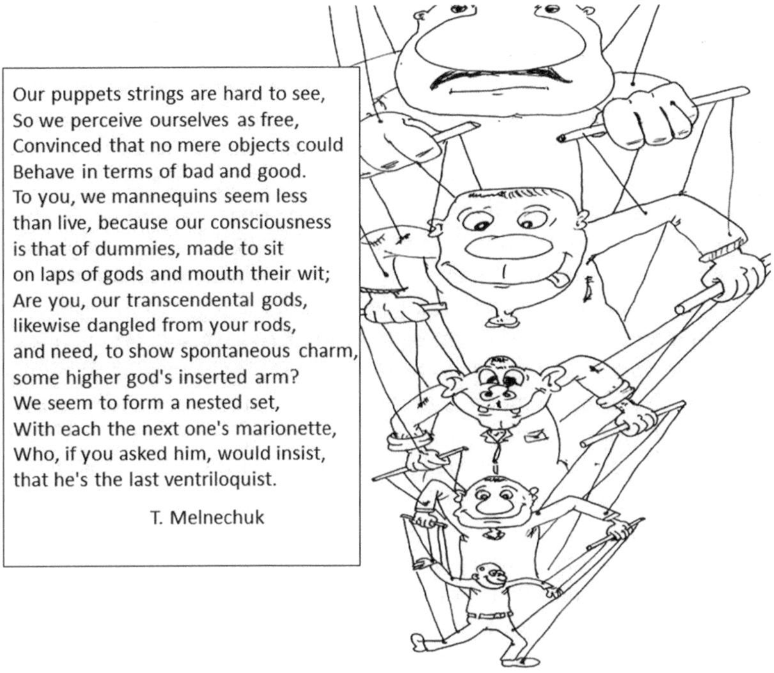

Fig. 2.5 Melnechuk's poem "Punch and Judy to their audience". Punch and Judy are two traditional puppets, and these words can easily be translated to the world of neuroscience, especially the matters treated in this section on whether there is a brain area in control of behaviour, whether there is a last ventriloquist in our heads handling the strings

to the feelings/perceptions of agency, volition and will. The results obtained with these patients present evidence for the importance of widespread activation of several brain regions in voluntary actions (see the Postscript to Chap. 2 for more on the importance of extensive neural activity for conscious volitional acts, when discussing alien hand syndrome) and revealed specific cortical areas that process information related to the intention to perform movements and the awareness of the motor actions, shedding some light on what possible coordination of activity between brain areas is necessary for the experience of intentional movements.

2.2.1 The Making of Choices: In Search of the Last Ventriloquist

So, how do we make choices? Is there any brain region more fundamental, perchance in charge? The evidence reviewed in the past sections already hints at some answer that there is not one main area in control; rather the "control of actions", as it were, is mediated by a widespread network of brain regions. Thus, we find areas in prefrontal and parietal cortices, in the brain's midline and even subcortical regions such as the basal ganglia. Whereas the neocortex and especially the motor-associated cortices that we saw previously (premotor and parietal areas) are more involved in planning, selecting and initiating actions to satisfy goals, it is the complex neuronal networks of the basal ganglia that are in charge of the gating mechanisms: the basal ganglia issue the final commands for movement executions when it is appropriate. And let us not forget the cerebellum, normally an ignored organ of the nervous system but essential for motor control—in fact, the cerebral and cerebellar areas are extensively connected via other brain regions like the thalamus and pons.

And what are the basal ganglia? While this is not the place to fully describe this very interesting region, the basal ganglia refer to a group of subcortical nuclei (that is, they lay deep underneath the brain surface or cortex) responsible for motor control, as well as other roles in motor learning, executive functions and even emotions (the basal ganglia neuroanatomy is reviewed in Lanciego et al., 2012). We will just say a few words about this brain region if only because the usual is to focus on cerebral areas like the prefrontal cortex when talking and explaining human behaviour, while these other neural circuits are somehow ignored, at least in non-technical texts, so here are some descriptions that will make you appreciate the importance of this region sitting deep into your brain. Did you know, for example, that the basal ganglia are a target to alleviate addiction? The basal ganglia and striatum are brain areas crucially involved in motivation and reinforcement behavioural phenomena, that is, pleasure. More specifically within these structures, the nucleus accumbens is a collection of cell networks considered to represent a limbic-motor interface mediating, in simple words, the neural processing of rewards, linking stimulus-reward association to behavioural outputs; without the accumbens, we would have problems enjoying food, friends and life in general. Consequently, the accumbens is related to the processes of addiction. In fact, work in our laboratory showed that reducing the excitability of the accumbens' neural masses by intracerebral injection of certain compounds reduced compulsive reward-seeking in rats (Kokarovtseva et al., 2009), and today, the accumbens is a target for deep brain stimulation to mitigate drug addiction, or addiction in general. Let us finish this basal ganglia excursion commenting that pathological changes in this structure, specifically the substantia nigra component, can result in syndromes like Parkinson's disease. This is mainly caused by a loss of dopaminergic neurons, and because dopamine is a major neurotransmitter mediating the processing of pleasure and reward, the treatment of Parkinson's disease with dopaminergic drugs can lead to addiction as a side effect. Now you know a bit more about the importance of deep brain regions in determining our habits and behavioural dispositions, it is not all in the neocortex.

We see then that basal ganglia and many other brain areas are in charge of processing information, of making us act. These are our ventriloquists. The concept of ventriloquist in this context is taken after the limerick written by the polymath Theodore Melnechuk which can be read in Fig. 2.5 along with my cartoon rendition, a poem that succinctly describes (and even solves!) the problem at hand. Readers who have gone over our text from the first pages will already suspect the picture that emerges. Putting together all the evidence obtained in brain research either in laboratory experiments or during observations of pathological syndromes, the scenario is that, just like the puppets try to express in that poem, there is no single puppeteer, or ventriloquist, fully in charge; rather, many regions of the nervous system (even beyond the brain) have moments where they can be considered the last ventriloquists, those that determine the behaviour. This conjecture will be further substantiated in the next section on the remarkable cases of split-brain patients.

To make a long neuroscientific story short, decisions emerge from highly distributed processes—manifested by organised or coordinated neuronal activity—in a variety of neural networks that mutually activate themselves. The brain regions specified in the first paragraph above are just a few of those thought to be involved in making choices. In truth, it could be said that almost the whole brain is implicated. Naturally, the question of directionality of activation arises: is it the activity in basal ganglia neurons that activate those in the frontal cortex, or is it the parietal and cerebellar neuronal networks that force others in basal ganglia and motor cortex into action? We will not go into establishing a specific directionality; this would belong to a technical text and not to this book. Nevertheless, we shall mention some intriguing observations that have generated much debate.

In the year 1983, a paper appeared with the following provocative words in the title: "the unconscious initiation of a freely voluntary act". How can we unconsciously determine a free-willed action? Aren't we supposed to be fully aware of the reasons to perform those actions? The paper described the very famous Libet experiment, where Benjamin Libet and colleagues demonstrated that there is a very specific electrical waveform recorded, using EEG, in some brain regions like the motor cortical areas—scalp EEG records cortical neuronal activity (well, and scalp muscle activity too, but let us skip this lest we become entangled in a very technical matter). This waveform occurs before the individual has the conscious intention to move his, say, arm—in the original experiment, the actions consisted of pressing a button or flexing a finger (Libet et al., 1983). This experiment spawned much controversy in the fields of consciousness research, free will, volition and the neurophysiology of agency. The controversy centred on the fact that if there is activity in your brain preparing an action of which you are unaware or have not consciously thought of yet, then this unconscious neuronal activity occurring in some parts of your brain will make these brain regions candidates to be the last "ventriloquist" talked about above. To be fair, Libet's findings are based on what Hans Helmut Kornhuber and Lüder Deecke published years before, in 1964. They reported that there are some unconscious electrical processes in the brain whose waveform was originally called the Bereitschaftspotential (readiness potential in English), or BP for short, which precedes conscious decisions to perform volitional, spontaneous acts. The waveform starts to become apparent ~1.5 seconds before movement execution. This implies that unconscious neuronal processes precede and potentially cause volitional acts which are retrospectively felt to be consciously motivated by the subject. Hence, the apparent conclusion from these experiments is that conscious intention arises after the initial stages of unconscious motor preparation. It is of note that the BP appears always related to the preparation and execution of voluntary movements, and only in exceptional cases does the BP appear preceding *involuntary* actions. For instance, it is absent prior to externally triggered actions such as a command from someone else to move your hand, and therefore, it is thought it represents a neural correlate of voluntary actions. While Kornhuber and Deecke found the BP to precede voluntary movements, it was the design of the experiments by Libet and associates that placed this brain waveform within the context of wilful actions by inspecting the relation between the time and shape of this electrical activity with the psychological, subjective feeling of the will to perform an action.

So this is the scenario deriving from these observations: you are now comfortably sitting and reading this amazing book, when suddenly you feel an urge to stand up and walk a few steps. If asked, you will swear that you wanted to stand up and walk and that, for sure, some neural activity could have been recorded (had you had an EEG cap on your head) occurring at the time of your conscious, volitional act of desiring to stand up. But thanks to Kornhuber, Deecke, Libet and colleagues, we know that about 1.5 seconds before you had the urge to close the book and walk, parts of your brain (motor cortex) were already displaying the neural activity—that BP—that will determine you walking (and stopping reading this most interesting book!). In fact, that subcortical region introduced above, the basal ganglia, would have shown electrical signals even earlier: 2 seconds prior to movement execution. So it seems that it is about 2 seconds prior to your "willed" standing up when some ventriloquists within your skull start voicing their opinion as to what forthcoming "voluntary" action you will be taking. But wait, is it 2 seconds, or can we go back even earlier in time? We have to consider that the neural activity recorded by the usual techniques like electroencephalography or intracerebral electrodes reflects not only the output of neurons but also the input to the cells in the neighbourhood of the electrode or the sensor. The output is those action potentials, or spikes, introduced in Sect. 1.4, and the inputs are the synaptic potentials which neurons receive from others connected to them. These latter potentials represent information coming from the connected neurons one step back in the chain of neural networks. In fact, for reasons that we will not go into, these recording methods reflect more input (synaptic potentials) than output, and the technical term—for those who may be interested to follow up—for what these sensors/electrodes record is field potential: the summed activity in many cells surrounding the sensor. Hence, considering this and the never-ending mutual activation of neural networks, that activity seen in basal ganglia neurons 2 seconds before movement reflects mostly input, and therefore, one may ask, from where are those signals arriving at the basal ganglia? It seems that there are other brain areas sending these synaptic inputs (information) that in turn will activate basal ganglia cells which will further activate motor areas in the cortex following those neural chains of activations, until you finally perform the action. But then we are in peril of stepping into an endless, closed loop, because the activity in those presumed brain regions that activate the basal ganglia also must be themselves activated, must receive information from other brain parts connected to them... You can see this is the start of an infinite loop from where we may end up right at the beginning, the motor cortex that moves the muscles via the spinal cord, since in the brain everything is connected to everything directly or indirectly. To avoid being trapped into the black hole of infinite loops, let us be content admitting that about 2 seconds before you consciously decide to act, there is brain activity planning that action.

To sum up, according to all the reflections reviewed in this chapter so far, it may not be feasible to localise the brain area where the neuronal activity initiates your impulse to perform an action or a choice. In fact, it may not even be one unique brain area; it could be various neural networks spread over the nervous system that are starting to induce the rest of the brain—to incorporate or recruit the motor cortical areas into that chain of activity—to perform such action. At some point, the neural networks activated are those belonging to an area where the perception of the urge to act is felt, and what area that may be? Neuroimaging studies and brain recordings in animals suggest that frontal areas generate the prior intention, and the perception of this intention occurs later perhaps in the parietal cortex; the evidence, however, is still unclear. The fundamental point is again that of neural chains of activations that recruit other neuronal networks into action. In order to fully understand this point, in Sect. 2.2.2, the case of an intriguing automatism during epileptic seizures in a patient is fully described because it reveals the basic nature of our actions.

This readiness potential occurs before voluntary actions, but is this true of voluntary thoughts? Does a readiness potential occur before a willed thought? Are there parts of our brains that also determine what we will be "voluntarily" musing about? At the time of this writing, we could not find anything in this regard to answer the question. However, very close to this is the fact that in cases of imagined (but not executed) movements, the BP also appears, although a bit reduced in amplitude compared to that of the moments when the movement was in fact carried out (Castro et al., 2005).

The controversy raised by the Libet experiments ranges from the field of philosophy and consciousness research to the more technical neurophysiological and psychological aspects of the experiments. Readers who are intrigued about this heated debate can find numerous articles, book chapters and talks on this topic, which may help comprehend the not-so-trivial nature of the Libet experiment, if only because at the moment a subjective response is required from the participants—in this case, when they perceived the urge to act—in a supposedly "objective" scientific experimental condition, this opens a can of worms (metaphorically speaking). As an aside, the very foundation of scientific research is the obtention of objective results from reproducible experiments in strictly controlled experimental situations, but in our opinion, this is, in many cases, a fallacy (reasons explained in Chapter 2.2 of *The Rise of the Scientist-Bureaucrat —Survival Guide for Researchers in the 21st Century*, published by Springer). In any event, we must be fair to Libet and colleagues, in spite of the shortcomings of the experiment and the interpretations they decided to act; they did an experiment addressing free will, rather than doing what was the usual around that time and before them: discussing ethereal opinions and becoming embroiled in philosophical arguments. Like it or not, their empirical observations were a beginning in this field of research.

Let us imagine that all this is true, that there are brain structures that have "decided" to perform an action before the neural events reach other brain areas where this decision becomes conscious, and one such area could conceivably be the prefrontal cortex (we have described before that this region is a major brain association area involved in the general regulation of the behaviour). The parietal cortex is another possible candidate (recall from the previous section the Fig. 2.4 and the neurostimulation experiments). All these are areas that when receiving those chains of neuronal activations cause you to retrospectively feel/perceive that you have consciously desired to cause the act. Does this mean we are not as free to act as we thought? But why do we need to be conscious of our decision-making to be "free"? Does reflection upon actions make them "free"? The fundamental issue here is that many behaviours escape our awareness, so many of our actions have a reflex-like character, and we are sorry to perhaps belabour this point but it is a most fundamental one in understanding the nature of volitional acts (see Fig. 2.6 for the brain's model of volition). First of all, many responses are simple habits maintained by the stimulus-response reinforcement mechanisms. Perhaps most of our behaviour occurs like a reflex, just think about yourself in everyday situations, when you make a grimace as soon as you see or hear something upsetting, or a smile when the opposite occurs. We each have a long list of daily actions which have the same semi-automatic nature. Second, the fact that we do not need to be aware of many actions has evolutionary advantages. If you are a wolf hunting with your pack, you want to be using your brain powers to coordinate your actions with the rest of the group rather than thinking about how to coordinate your hind and front legs to start a run. We should not be concerned with being aware of the intricacies of coordinating the immense variety of musculoskeletal adjustments needed to translate our intention into actions. Indeed, scholars like Bjorn Merker have proposed that consciousness arose as an interface between the spatial senses and the motor requirements of motivated behaviour; in his own words "consciousness arose as a comprehensive solution to the logistic problems created by self-motion in one such domain of central control" (Merker, 2005). Looking back and learning from animals, our nervous system in these particular aspects may be a natural progression over those of other animals. Many animals do not rely on a centralised neural area to control the behaviour, as anybody who has cut in half a worm or insect has found out the body still survives separated from their "brains". And also recall from Sect. 1.7 that octopuses have a pseudoautonomous control of locomotor activity that resides in neural networks within the tentacles and not so much in their brains (or central ganglia in their case). So it is not that bad if the brain deceives us (that is, deceives itself) into thinking that our prefrontal cortex, or whatever other areas for conscious action control may be, has willed to commit a specific action based on a more or less logical chain of reasoning. In the final analysis, as M. Gazzaniga said, "The person is a conglomerate of independently functioning mental systems that in the main reflect nonverbal processing systems in the brain" (Gazzaniga, 1980).

Can then the concept of a completely unbiased, fully conscious and nondeterministic free will be saved? Some accounts emphasise the random, unconstrained and unpredictable nature of self-initiated actions, but we would advise caution when

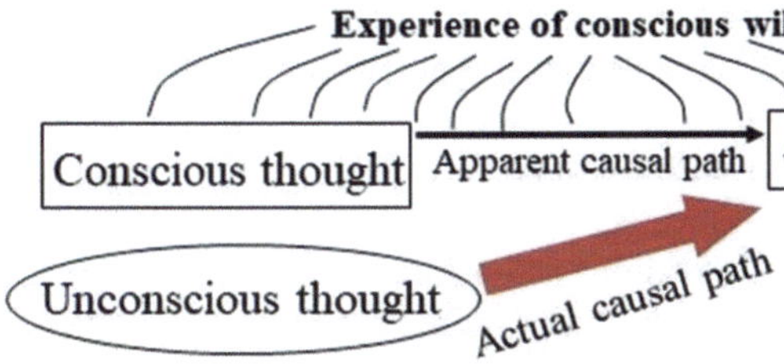

Fig. 2.6 The perception of freedom in our choices. Many brain processes are unconscious; that is what Gazzaniga's words about non-verbal processing shown in the text refer to, since our logical, conscious thinking normally is "verbalised", though sometimes mentally without explicit vocalisations. But it is only the conscious thoughts that we are aware of; hence, an apparent causal path for our choices/actions is established, although much of the bias in those choices remain out of our awareness. Between the conscious and the unconscious processing, an action is selected to be performed, such that the actual causal path indicated in the figure as the thick arrow is in reality the combination of both. See text for the enjoyable advantages of all these abundant unconscious brain mechanisms. As David Hume said in his 1739 'Treatise on Human Nature', "the constant union and inference of the mind that establishes causality in physical events must also give rise to causality in actions of the mind"

using these notions of randomness or unpredictability because these are relative concepts, as will be revealed in Sect. 8.1.3 in Part II. Nevertheless, there are some who invoke quantum indeterminacy in order to save free will, at least in the sense that not all in our will is determined by unconscious or partly conscious neural events in the brain; rather, there is some inherent randomness in the neural activity stemming from this subatomic indeterminacy. Short technical note: this quantum indeterminacy comes from the celebrated Heisenberg's uncertainty principle of 1927, stating that the position and the velocity of an object—atom or elementary particle in effect—cannot both be exactly measured at the same time. There is no need to go now into the issue of the possible application of quantum physics to the realm of consciousness, starting with the fact that the quantum scale is extremely short in time and space (we are talking about femtoseconds and nanometres), whereas at the psychological level the scales are much different—at least the time scale, where we can talk about seconds or maybe a fraction of a second but not femtoseconds (one needs at least a fraction of a second to be aware of some cogitation) and as to the spatial scale that would apply to psychology, perhaps metres or centimetres in our relations to others and other objects.

Nonetheless, there is abundant literature on the topic; a concise recent review on quantum tales applied to the neuroscience of consciousness can be found in Guevara et al., 2020. But even if it was possible to accept that the randomness of the quantum indeterminacy implies unpredictability of our actions, this adds little to the conscious free will notion because, if it is random, then where is the presumed chain of logical and determinate reasoning that makes us choose? Aren't conscious choices supposed to be determined by our deterministic (yes, a bit redundant) thoughts? Volition, after all, is the cognitive process by which an individual decides on and commits to a particular course of action, therefore not much randomness should be implicated. This notion of randomness in the neuronal mechanisms that determine behaviours therefore does not seem to do much to save the free will concept. So conceivably the philosopher and physician John Locke was right when he separated free from will, in 1690 he said, "I think the question is not proper whether the will be free, but whether a man be free". Writers and scientists alike have had common thoughts on the topic, putting emphasis on the choosing, rather than on the freedom, for instance, Paulo Coelho in '*O Zahir*' said, "Freedom is not the absence of compromise, but the capacity to choose", which parallels the previously shown Schall's thoughts that what exists is our capacity to choose but the choices are constrained. And we can add that it is nearly impossible to be consciously aware of all those constraints that in the end result in one of our choices. So when you choose one wine over another tonight at the restaurant, be content with the fact that you can choose and enjoy the wine.

2.2.2 The Will in Epileptic Patients: What Seizures Reveal About Volition

Patterns of neural activity manifest themselves as behaviours and—it could also be said in order to avoid dualism and dichotomies—behaviours are the neural patterns, such that brain activity and behaviour are two sides of the same coin. In the past section, we have searched for an ultimate ventriloquist, a brain region that may be in full control of our actions, and we have seen that the concept of central control may be a fallacy. The scientific observations reviewed above indicate that it is almost the whole brain that is involved in performing an action. As the neurologist Mark Hallett claimed: "At any one time, the activity of the motor cortex […] will reflect virtually all the activity in the entire brain. […] Is it necessary that there be anything else? This can be a complete description of the process of movement selection, and even if there is something more—like free will—it would have to operate through such neuronal mechanisms" (Hallett, 2007). As a perfect illustration of the phenomenon of how brain areas recruit others into action performance, sometimes consciously and other times unconsciously, we will describe here a remarkable automatism we saw in a patient during her epileptic ictal events (seizures).

Some epileptic patients have automatisms during their seizures. Behavioural motor automatisms are in fact a frequent ictal and postictal correlate of focal seizures. In some cases, for example, you can see the patient pick up a beer can and execute movements as if he were to drink the contents, but all these actions happening during the ictal event hence out of his awareness: this is why these behaviours are called ictal automatisms. For those who think of ictal events as being associated with tremendous uncontrollable convulsions, this may be surprising, so we note that not all seizures are convulsive, especially focal seizures where the intense neural activity occurs in a localised brain area and does not spread throughout the whole brain. Hence, in focal seizures, there are not big paroxysms, but that localised neural activity is enough to trigger the automatisms. And even in those seizures where patients suffer convulsions, there may be a quiet period towards the end of the ictus. And this has been our extremely brief epilepsy teaching session, those interested in this topic can find an enormous literature, and we offer a basic review on epilepsies in Perez Velazquez & Wennberg, 2004. For its particular interest, we shall describe the case of a patient who made, unconsciously, the sign of the cross during her seizures, ictal events which were documented by intracranial depth electrode and simultaneous scalp video-EEG recordings.

It so happened that, towards the end of her (temporal lobe) seizures, the patient made the sign of the cross and immediately after also made a praying gesture (put the palm of her hands together). When told that she had performed the sign of the cross during the seizures, she had no knowledge that it occurred, but she spontaneously offered that she had been raised as a strict Catholic. While this could be one random automatism of those found in some patients during ictal events, in her case, there was a suggestion that making the praying gestures may have represented a learned ictal behavioural phenomenon Or, in simpler words, that the seizure caused a behaviour she learned in her past to manifest itself during the abnormal neural activity in the ictus. The reason we proposed this hypothesis is that she had been forced in her youth, over a period of many years, to make these gestures in the post-ictal period (immediately after the seizure had finished) as an atonement, that is, when her seizures had finished her parents made her cross herself and pray—we have to acknowledge that even in this time and age, there are people who believe that ictal episodes are manifestations of possessions. The possibility is that these movements, which in early times were performed consciously immediately after her seizures, ultimately came to be performed unconsciously, during the ictus, associated with the neuronal seizure discharge in temporal lobe neural networks. Hence, it looks like the neural activity during seizures was able to recruit and activate an adjacent neural memory circuit storing the praying gestures (Wennberg et al., 2009).

Why is this, as well as other cases of automatisms during seizures, relevant for our purposes here? Because these cases illustrate the fact that our behaviour is the result of those unremitting and never-ending neuronal chains and that whether some conscious control is exercised or not will depend on which neural networks or brain regions become active. In terms of the neurophysiology, or brain dynamics, the phenomenon just described of that patient performing praying gestures is nearly identical when she was young and forced by her parents to consciously perform the actions and later on when the signs were done unconsciously by the recruitment of those neural networks where the movements were stored, this recruitment due to the high neural activity during the ictus. Now extrapolate these considerations to other human actions, if only to understand better how people behave or misbehave. Imagine that someone is being nasty to you. You talk to your bully, trying to make him reason, to analyse his action and to recognise that his behaviour is unjust towards you. However, if the bully's "reasoning" neural networks are not more powerful than those in charge of the bad behaviours, then all that reasoning will be futile. The usual belief most people have is that there is some reasoning that can be explained to "bad people" so that they amend their misbehaviours. This in neuroscientific terms means that we think there are brain areas which may block or veto the bad action. Given the neuroscientific evidence that we have presented thus far, this is not necessarily true; it all depends on how strong those neural networks are in terms of forcing other networks to become active. Is the bully "programmed" in his behaviour? Let us examine another scenario. Imagine that you are a smoker and are considering your next cigarette. Whether or not you will end up smoking this cigarette will depend on how many neural networks are active. You may have competing thoughts: you have had "too many" that day, smoking causes cancer, nicotine feels good, and the costs of smokes. The relative strength of these competing networks will determine which will activate the final neural networks that will cause either the hand to pick up the cigarette and light it or to forget about it. And some of these neural chains of activity will be conscious and others not.

So some criminals—say, a serial offender like the typical serial killer—have little choice but being criminals, the brain areas involved in being violent and performing crimes are just too powerful and control the motor cortex that initiates that person's actions. Some notorious offenders truly realise this; they know that their "good" neural networks that could stop or change the activity of the "bad" networks are just too weak, and as a result, some of these criminals either commit or try to commit suicide, this showing the great extent that they are aware of their situation. Therefore, this understanding of the basic neurophysiology and brain dynamics associated with making choices and behaving in general help us understand better the predicament of those who misbehave. But this does not mean we have to love and nurture them; those who are dangerous have to be isolated where they cannot inflict pain to others. Nor do we have to hate them, but rather understand that this is how their brains work. Just as someone can develop a kidney or heart disease, brains too can develop pathological syndromes. The big difference, the reason why nobody will blame you for acquiring a kidney disease but they will blame you for being a bad person, is that kidneys do not determine behaviours. Or do they? Remember that everything in the body is interconnected, so renal dysfunction leads to inefficient removal of waste from the body, such that the accumulation of toxins and high levels of urea will affect brain function, causing confusion. Similarly, high blood sugar levels in patients with diabetes can lead to delirium and psychosis. Still, we would not blame the person for their chronic kidney disease or diabetes.

We can add that a similar plight is suffered by addicts. Although we all have our little addictions, it is in the pathological cases—the real addicts—where their neural areas involved in processing rewards (recall the nucleus accumbens aforementioned) are just too strong and powerfully activate other neural networks resulting in compulsive reward-seeking. The book *The Man Beside* by Steve Campion offers a unique account of one of these addicts and explains why an addict will always remain an addict. Such is the stability of the brain dynamics that determine reward-seeking, for after all this is the basis of animal behaviour. Like in the case of the serial offenders aforementioned, some addicts realise their predicament. Homeless youth addicted to street drugs from crystal meth to heroin, when interviewed, would say that they knew the dangers of their drug and possible complications—HIV, infection and death—speaking dispassionately, without fear of the complications.

This would underscore the point of a certain brain network having precedence and driving the behaviour, acting as the last ventriloquist.

Brains, then, are viewed from a different perspective than other organs, but it should not be so: brains become ill and will produce "bad" behaviour like a sick stomach will produce bad digestion; it is that simple. Although it is true that some brain disorders are more normally considered as pathologies, for instance, nobody would say to a Parkinsonian sufferer, "Quit having tremors. You are making me nervous", or to an epileptic patient, "Stop convulsing. Don't you know how to behave properly". So next time you tell an addict "Stop taking those substances because they are killing you", think that it may not be as easy as what you, with a non-pathological brain, could achieve. Understanding how brains function helps understand us; hopefully, this book (and others more technical, e.g. Perez Velazquez & Frantseva, 2009) may reveal something useful in this regard so that mutual understanding of our multiple behaviours, pathological and not pathological, spreads over the planet, something badly needed if we are to live as a society.

The epileptic case described in this section demonstrates too what we have been expressing in some parts of Chap. 2 (recall echopraxia, utilisation behaviour, etc.)—that some of our behaviours are stored in certain brain cell networks and the nature of the reflex-like actions we sometimes perform. In the case above, the network for the sign of the cross was "activated", so to speak, by the ictus, but in more mundane situations, there is no need for a seizure to make that network come alive. If you had a Christian upbringing, you have probably witnessed people cross themselves in response to seeing or hearing something nasty or unpleasant. In this situation, it is not an ictus that recruited the sign-of-cross cell network; it was another or several others brain areas connected to that one, and that became active. This "reflex" response has been conditioned over time and socially reinforced. As mentioned above in Sect. 2.2 in the case of electrical stimulation of brain regions that cause specific behaviours, we see now more evidence that in the end, it is all how those brain cell networks establish the chains of activity, and it matters little whether it is artificial neurostimulation with electrodes, abnormal epileptic activity in seizures or mundane situations with normal neurophysiology underlying the behaviour.

As a corollary, if all this is true, we could ask, like Hallett, "If there is no free will as a driving force, are persons responsible for their behaviour? This appears to be a difficult question, but it is not. It is difficult only for the dualist" (Hallett, 2007). Related to this issue of responsibility, the words of Michael Gazzaniga explain some basic truths: "Personal responsibility is a public concept. It exists in a group, not in an individual. If you were the only person on earth, there would be no concept of personal responsibility. Responsibility is a concept you have about other people's actions and they about yours. Brains are determined; people follow rules when they live together, and out of that interaction arises the concept of freedom of action [...]. They [*referring to these concepts of responsibility*] exist only in the relationships that exist when automatic brains interact with other automatic brains" (Gazzaniga, 2005). Note he says, "Brains are determined", so we are not the only ones telling you these ideas about the determinism in the neural chains of activity that results in behaviours; we are in good company. One can already sense that neuroscientific research in this field may be saying something about our legal system in regard to certain criminals and socially pathological behaviour. We will devote Chap. 17 in Part III to bring up the issue of how brain research is changing the legal system, especially criminal justice.

Let us end this section with a famous Dostoyevsky's paragraph on these topics that appeared in his 1864 novel *Notes from the Underground* (you can agree or disagree; it is all fine, as one purpose of the scientific discourse is to discuss and debate ideas): "Science will teach man that he never has really had any caprice or will of his own, and that he himself is something in the nature of a piano key…and that there are, besides, things called the laws of nature; so that everything he does is not done by his willing it, but is done of itself, by the laws of nature. Consequently, we have only to discover these laws of nature, and man will no longer have to answer for his actions, and life will become exceedingly easy for him."

In this chapter, we have introduced some concepts that could be called the free will's retinue, particularly agency—the feeling of being the agent of one's thoughts and actions—, but this implies there should be one or more brain regions that interpret those actions and attribute agency. This presumed interpreter has to assign value to, or evaluate, the models brains make of the personal identity and the world around. There should then be another important member of that retinue, the interpreter within the brain; we will search for it in the next section reviewing the amazing world of split-brain patients.

Postscript to Chapter 2

To avoid excessive technicalities, we have not had the opportunity to discuss several neuropathologies that further illustrate the complications the brain encounters in order to interpret our presumed willed actions. For example, one could talk about lesions to the frontal lobe resulting in the lack of volition and long-range planning. But due to their singularity, it could be more interesting and instructive to consider the alien hand syndromes, when patients perform movements without conscious will; normally, it is one hand that is not under the control of the individual, and while it performs goal-directed tasks and actions, the patient will claim it is not him/her who directs that limb. If you have seen the actor Peter Sellers as Dr. Strangelove in the 1964 movie of the same title, you know what this is. This condition is associated with injuries like cerebral stroke, the presence of tumours or surgical procedures like callosotomy (described in the following section on split-brain patients), and among the main brain areas thought to be involved, we find our two acquaintances we keep meeting: frontal and parietal lobes. And if we are allowed a more specialised comment that supports the importance of neural organised activity that we have mentioned many times as fundamental for healthy brain function, it is of interest that in brain recordings using neuroimaging methods during the episodes of alien hand movements, it was detected that the neural coordinated activity differs from that associated with the movements when the actions are voluntary. So, the alien movements had more localised brain activations (around cortical motor areas like M1) as opposed to voluntary motions where several other areas showed activity such as the premotor cortex and frontal gyrus (Assal et al., 2007). This observation is important because it indicates that for the brain to "understand" or interpret correctly an action, the neural activity has to spread over several regions, especially those like frontal areas involved in goal-directed actions, so the more neural networks active, the easier to be aware and organise the willed actions (the neurostimulation experiments described in Sect. 2.2 support this notion).

Or we could talk about the neglect syndromes, when sufferers experience the denial of ownership of body parts or ignore parts of the surroundings as if these did not exist at all. After injuries like stroke, there is extensive cell death in some brain areas that causes the breakdown of the functional connectivity in frontal and parietal networks (yes, again these two lobes). Patients will fail to explore or notice the affected side or react to stimuli located on this site (recall that the left hemisphere is in charge of the right side of the body and the right hemisphere controls the left side). In the syndrome called asomatognosia, for example, there is a disturbance in the awareness of one's own body, such that the patient will fail to recognise or identify a specific part of the body, usually a limb or part of a limb.

And to end this postscript, something that can be considered fascinating: how do you cure neglect syndromes? Well, besides sophisticated neurostimulation methods (as could be expected: if there is a problem with brain activity, then try to alter that activity to improve the condition, and neurostimulation is the most direct and specific procedure to change neural activity), there is something tremendously simple: caloric stimulation. Remember the vestibular system housed in the ears (Sect. 1.8.1). The therapy consists of irrigating the auditory canal with cold water. In simple words, put ice-cold water into your ear, and within a few seconds, you will start to recognise your, say, right arm as being yours (in case you suffered asomatognosia), or you will start paying attention to the right-hand side of your field of vision (in case you had spatial neglect)—but please do not try this at home without medical advice. This effect is transient and lasts for a few minutes. This remarkably easy technique transiently cures neglect syndromes, phantom limb pain (Sects. 1.8 and 2.1), anosognosia (denial of deficits), somatoparaphrenia (bizarre beliefs regarding hemiplegic limbs), bipolar disorders and chronic pain, and if all this were not enough, it has been used as well to modulate a wide range of cognitive and sensory functions in brain-damaged patients and in healthy participants. Not bad for such a simple procedure. But, why these remarkable effects of just cooling the ears? Recall that the vestibular organs encode things like body position, perception of self-motion and the connections to the brainstem help bodily self-consciousness. The cold water in contact with the external auditory canal induces a change in temperature that leads to convection currents in the semicircular canals, and, in simple terms, this causes neuronal activations that spread to almost all cortical areas, especially in the parietal lobe; thus, this vestibular stimulation produces a sort of "non-specific" responses in most brain cortical regions, and it is thought that the transient reversal of symptoms occurs by activating thalamo-cortical activity that reintegrates impaired cortical regions that are operating in a sort of disconnected manner in those patients.

Quotation marks were used above in the word nonspecific because it is a very ambiguous term when applied to complex systems like brains, and although scientists love to talk about "specific" mechanisms of this or that drug, protein or therapy, when studying complex nonlinear phenomena, specificity is restricted to a small localised domain in space and to a very short time scale. Being nonspecific sometimes has advantages, like in this case of caloric vestibular therapy or in other protocols that consist in activating many brain areas in order to, for instance, awaken patients from the minimally conscious state (a condition similar to the vegetative state but not as severe). The logic behind the effectiveness of a wide nonspecific activation of the brain neural ensembles is that hopefully such over-activation of many neural networks may restore the normal function and connectivity among some of them which brings back normal neurophysiology and alleviates the syndromes. This could have been the reason why a patient who had been in a minimally conscious state for 6 years recovered alertness after deep-brain stimulation of the unspecific thalamo-cortical system through certain midline thalamic nuclei (case reported in Schiff et al., 2007). Therefore, perchance there should not be too much concern with being nonspecific, in cases where specificity cannot be achieved.

Chapter 3
The Strange World of Split-brain Patients: In Search of the Interpreter of Our Actions

Contents

Along the pages of this book, we have seen that fooling our perception is not that difficult. In the preceding chapter, we saw how people can be fooled into thinking they performed a movement—even though they did not—by stimulating certain brain areas like the parietal lobe (Fig. 2.4), or thinking they are elsewhere out of their bodies by neurostimulation of the TPJ that "sends" them out of their bodies (Fig. 2.3), or believing a rubber hand is their real hand by patting the hand and misleading their sight at the same time (Fig. 2.1). In sum, all these experiments that involve neurostimulation or other types of sensory stimulation evoking conscious intentions, unconscious or illusory movements, suggest that there is a brain area interpreting the action, or the lack of it. Brains make models of the environment and of the self, and along with it its features such as the agency underlying volition, for which the individual has to interpret the behaviour. If, as we saw in the previous section, there may be several ventriloquists within our skull vying for power to control our actions and choices, the question is whether there is as well some brain area that acts as the interpreter of those actions determined by the, perhaps the last, ventriloquist. The split-brain world sheds light on this matter.

The term "split-brain patients" refers to those who have had their corpus callosum sectioned, either partially or totally. The surgery is called callosotomy, and it is used to treat epilepsy in an attempt to prevent the spread of epileptic seizures because the callosum connects both brain hemispheres. In simple terms, it is a bunch of nerves (the axons of neurons) that cross the brain midline and connects symmetrical cortical areas. There are other nerves connecting the two hemispheres, called commissures, but the callosum is the largest, containing about two hundred million axons. If you are into numbers, brain numbers are staggering: from the

J. L. Perez Velazquez, V. Nenadovic, *Being and becoming*,
https://doi.org/10.1007/978-3-030-78264-1_3

number of cells in the brain to the number of synaptic contacts, quantities rival those of stars in the sky (see Chap. 6 where some of these astronomical magnitudes are shown).

Many readers will have come across the reports about the left hemisphere being important for certain things (e.g. logic, language) and the right for other aspects of the behaviour (e.g., creativity, emotions). Could it be that one hemisphere, perhaps the logical left hemisphere, is the interpreter of our behaviours? We will see about this in the following paragraphs, but one thing that we should clarify now is that the right hemisphere controls the left side of the body, and the left hemisphere the right side.

It is always of interest to see how ideas about the nervous system evolved. Thus, in the not-so-distant past, the corpus callosum was thought to be the site of the soul. This idea started around the 1730s and, more recently in the early twentieth century, the great psychologist Karl Lashley proposed that its main function is to prevent the hemispheres from collapsing onto each other. Oh well, now we know a bit more: the direct communication it offers between the two brain hemispheres is essential for a comprehensive view of the surroundings and our actions on our environment. In this manner, and to grossly simplify the truth, when we move both our hands to catch a bird, the two limbs are really working in a coordinated way, but if the hemispheres were totally disconnected, then the right brain would move the left hand, and the left brain would move the right hand almost independently which would pose problems in our catch.

In the early days of split brains, this term was first used to describe the surgical procedure performed on cats and monkeys, in experiments trying to study the visual pathways for right and left integration. This work was pioneered by Roger Sperry (we met him in Sect. 1.2), who perfected the operation in animals. In the 1940s it was used on humans—as mentioned above the purpose was to prevent epileptic seizures to spread to the other brain hemisphere as the callosum connects both hemispheres. The results were so spectacular and important not only as a therapy against seizures but also for what was learnt about brains, that Sperry received the Nobel Prize, 20 years after the first split-brain operation (you do have to wait for this type of prize!). What was of interest in the early days of the surgeries is that the patients' behaviour was essentially normal, in contrast to what was reported in animal experiments where it was found that some information did not transfer between hemispheres. But this difference between the human and animal results soon vanished, when patients were probed in-depth, just like the animals had been in rigorous experimental conditions. It was Michael S. Gazzaniga, among others, who meticulously studied some split-brain patients, like the famous patient W. J., observations described in his book *The Social Brain*. These observations in patients showed that, just like in cats and monkeys, the information did not transfer between the patient's hemispheres. In a few simple words, in everyday life, these patients can manage very well because the right and left brains, while separated, have equal access to the information in front of them, but interesting things began to appear when experimental conditions did not allow, say, the left hemisphere to see what the right side was seeing. The preliminary interpretation of these findings was that there was a dissociation of competencies between the two half-brains.

Then, it can be expected that in some specific matters these patients will be in difficulty. For example, in Fig. 3.1, it can be seen how one patient was able to draw a cube with both of his hands, but after the callosotomy separated his cerebral hemispheres, the drawing executed by the right hand was poorer than that of the left hand; this occurred because the left hand is mainly controlled by the right hemisphere that is good at processing spatial shapes, while the left hemisphere, which controls the right hand, is more concerned with language and logical thought, and not so good at artistry.

Now, watching those drawings in the figure one can wonder whether the poor pictorial ability exhibited by the right hand reflects a deficit—of the left brain which controls that hand— either in the execution or in the perception of the task, or in both. Admittedly, this seems like trying to separate two aspects, perception and action, that in reality should never be separated because one comes with the other: when we act, we perceive the consequences of that action and we adjust the movements or whatever is needed. Nonetheless, we scientists must admit this attempt at the partition of the mechanisms observed in natural phenomena is a sin we commit on many occasions, and those scientists much into the reductionist paradigm—the view that to study a complex phenomenon the best is to split it into separate entities and studying their independent functioning—sin even more. Be as it may, sometimes it may be worth asking these questions, so as an attempt to address the aforesaid question, it so happened that when these patients were requested to manually arrange blocks to match a certain configuration, the same differences between right

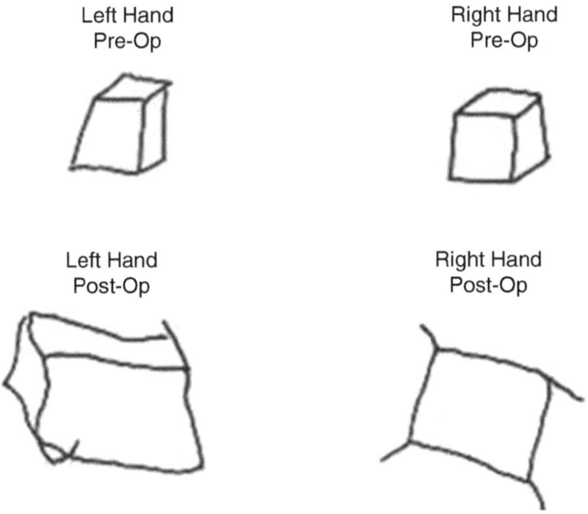

Fig. 3.1 One patient's drawings of a cube with the left and right hand, before the callosotomy (above, pre-Op), and after the surgery (post-Op). (Adapted from M. Gazzaniga's *The Social Brain* (Basic Books Inc. 1985)

and left hands were observed, but if they are asked to mentally do the arrangement without manipulation (for which they would answer by matching picture to picture) no differences are observed. In conclusion, it appears to be a matter of manipulation: the right brain seemed superior at manipulating objects using the left hand. This is an example of one asymmetry that was found in these individuals, suggesting that specific brain systems handle specific tasks; in other words, there is certain modularity in the brain neural circuits. This notion of brain modules, or areas specific for processing particular information, we already saw in Sect. 1.4 when it was mentioned that there are brain areas that process specific stimuli and sensorimotor transformations (auditory cortex for hearing, sensory cortex for touch), but recall that it was stated too that such segregation of regions is not at all strict, in that they are not completely anatomically and functionally independent.

But the observations on the split-brain patients that bear on the interpreter question are the following. A typical experiment goes like this (and remember that due to the callosotomy the right and left-brain hemispheres do not share information, so what one side sees is not transferred to the other side): a patient sits in front of a screen placed very near her body such that it prevents her watching what her hands will do, and on the screen two words are shown, each directed to one brain hemisphere, for example, "key" to the right hemisphere and "ring" to the left hemisphere. Technical note: how can you send different images to the two hemispheres? It is not as simple as showing one image to the left eye and the other image to the right eye because what is split up is the right and left visual fields and not the input from each eye, hence one needs a specific setup to project the images in front of the volunteers such that each visual field remains separated and thus the right field goes to the left brain and vice versa, and this ends our technical note. Then after being shown the words, you ask the person to pick up the object she saw from a bunch of objects placed behind the screen, so using her tactile sensation she will pick up a key with her left hand (right hemisphere saw the word "key") and a ring with her right hand because this is the word the left brain saw, but if you ask her to verbalise what she grabbed, she will only reply "a ring". She is unable to say "a key" because the right brain is normally mute in people: the left hemisphere is where language resides, and hence only the left hemisphere can talk (this is true in most cases but as always happens there are some exceptions and there have been some patients who had a talking right hemisphere too). Of course, if the screen does not prevent her from watching her hands, she will have no problem declaring she picked up keys and rings because her talking left side could see what the left hand—driven by the right brain—did, so it the end it will be the same thing, the left brain doing the talking. But because each hand behaved properly picking up the right thing, this clearly demonstrates that each brain region—hemisphere—perceives the thing it is supposed to perceive (a word in this case) and accurately directs the behaviour of the corresponding body parts.

Thus, the right hemisphere is perfectly aware of what it perceived, it understands language (because a word was flashed to it on the screen and understood what it meant), but it just cannot verbalise it because it lacks the neural networks for processing language. However, the mute right brain is perfectly able to answer if asked in the appropriate manner, in the experiment it answered by directing the hand to pick up a key. In an effort to put things together in a coherent perspective—something scientists must always try to do—recall that when we talked about subliminal perception and when we described other things like the neurostimulation experiments of Sect. 2.2, it was already pointed out that each brain region perceives what it is supposed to perceive, but whether these perceptions can have access to the neuronal networks responsible for conscious awareness or whether they can be verbalised—or communicated in some ways—is another matter. And if you want one more striking example of this, but not from the split-brain world, is that of patient D.F. who, after brain damage, was very poor at describing or demonstrating the orientation of a line or slot, and yet she could still reach out and post a card into the same slot without error (Milner et al., 1991). So here we have another striking example on the matter of reportable mental events: certain brain networks in D.F.'s brain were perfectly aware of the orientation and shape of the slot and could direct a hand to post a card, but could not verbalise, could not describe the orientation of that slot because the brain areas needed to achieve such verbalisation were damaged. In technical parlance, she had a dissociation between phenomenal conscious content and visuomotor transformations. These reports should prompt us to seriously consider the implications for our behaviours and judgements of the fact that parts of the brain, but not all the brain, can be aware of particular mental states or perceptions and how these parts can force us to act or judge.

Going back to the split brains, that typical experiment described above already suggests there is an area that acts as an interpreter of reality, in this case, it would be the brain left hemisphere. One more clear example of how brain networks are perfectly aware of things they perceive but yet cannot communicate unless properly asked is that of patient J.W., who after his right hemisphere was shown the word "bike" and was asked what he saw, he replied, "Didn't see anything"; this was obviously his talking left brain responding and because the left hemisphere saw nothing, then it is true, but when prompted to draw with his left hand (commanded by the right side that saw the word) he draws a bike. This drawing exercise took some convincing to do from the researcher because, naturally, J.W. protested that it was a ridiculous thing to try to draw the word he saw since he (the left brain) did not see any word. If asked why he drew the bike or whatever his right brain saw, he would normally say something like "I have no clue, what is it? It looks like [here the word he—the right brain—saw in each different experiment]". Hence the right hemisphere understands written language but cannot verbalise things. Of course, there is variability among individuals such that other split-brain patients have right hemispheres with almost no semantic capabilities, and the advantage of using these patients who had right hemispheres with enough semantic competence is that these experiments to assess how we interpret reality can be carried out in a most striking fashion. These and the following observations on other patients are fully described in Gazzaniga's aforementioned book.

What can be apparent now from these various observations performed on these patients we have described so far is that we have a sort of reportability problem: individuals, that is their brains, have to be asked appropriately, the response requirements which normally involve a motor action—either moving the mouth to talk or picking up an object or pointing at something, etc.—have to be of the right type so that the brain part that sensed some stimulus can express it. It is not all in the talking, as we have seen already in cases D.F. and J.W. These studies also say something about our everyday lives, because while we have a well-connected brain with an intact callosum, still same features are present like the ability of the left brain to speak and handle language-derived logical thoughts; so when we talk and express our opinion, do we really know what part of the brain is doing it, what neural networks that may be processing ideas and perceptions which remain unconscious to the left brain could be forcing that left brain talking area to express the idea or thought? Things to ponder about.

Let us describe another extremely interesting experiment to realise how far the brain, or the interpreter in the brain, can go in terms of confabulation. The experiment is described in Gazzaniga & LeDoux, 1978. A distinct image is shown to each hemisphere of a split-brain patient: the left side sees a bird's claw and the right sees a snow scene. The patient is told these images represent a "problem" that has to be solved, so after showing these images the patient is asked to choose a card, from a bunch in front of him, that serves as a possible answer to solve that "problem". The patient then chooses the card with a picture of a chicken with his right hand, which is fine because the left side moving the right hand saw a claw, and he chooses a card with a shovel with his left hand, which is also fine as the right side saw a snow scene so the shovel will fix the "problem". The interesting thing happened when the researchers asked him why he chose those things. In his own words: "that's easy, the bird claw goes with the chicken and you need a shovel to clean out the chicken shed". Here was the left brain explaining, or rather confabulating, why the left hand picked a shovel when the only image the left brain saw was a claw, thus in the end the patient's conscious story was along the lines that "I will use the shovel to remove hen's excrements", a story made up by the interpretation of the events by the left brain. There were many other experiments that demonstrated the ability to confabulate by the interpreter, normally the dominant left hemisphere. As another illustration, flashing the command "walk" to the mute right hemisphere caused the patient to stand up and walk, and when asked where he/she was going, the answer was of the sort of "going to my house to get some food". We see then the great power that some brain areas exercise on our behaviour and how the left brain tries hard to interpret those actions, being faced with the task of explaining the overt behaviour initiated by the disconnected right brain. The remarkable observations that instruct us about the conscious mechanisms we use to build up the sense of personal identity, that is self-awareness. Figure 3.2 is a sketch summarising the main concept derived from the split-brain cases.

Incidentally, there is what can be considered a natural split-brain model, it is called callosal agenesis: people born without the callosum, a condition first reported in 1812. These individuals are mostly normal in terms of IQ (that anthropocentric measure of intelligence) or performance of tasks in everyday life, but when probed in similar experiments as those done on the surgical split brains, the disconnection symptoms are comparable in nature but not in magnitude to those observed in

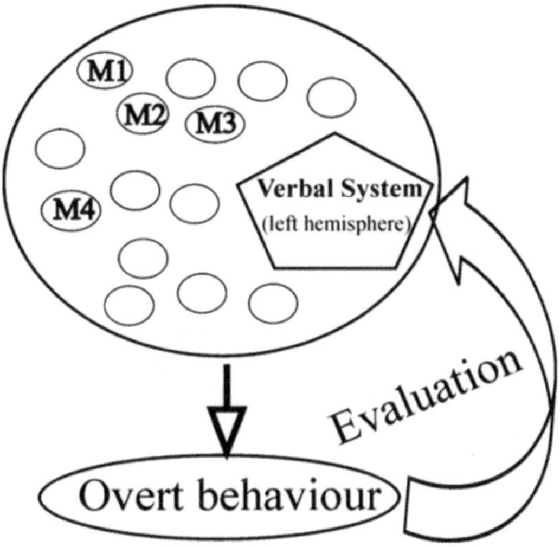

Fig. 3.2 Scenario that emerges from the observations on split-brain patients. Each small circle is a brain module or neural network in more neurophysiological terms, that can process a certain type of information (M1 may process auditory information, M2 visual, M3 somatosensory, etc.). The actions of all together result in our behaviour which is perceived by the brain and evaluated, or interpreted, by the dominant brain hemisphere, normally the left one where the language areas reside, although this does not imply it causes the behaviour. Hence this hemisphere could be the "last ventriloquist" we were searching for but not in terms of determining behaviours rather as a verbaliser and interpreter of those. It is in this sense that "Your conscious life is an 'afterthought' constructed by the interpreter", as Gazzaniga expressed in *Mind Matters: How Mind and Brain Interact to Create Our Conscious Lives*

patients. The fact that these symptoms are lessened in people born with the hemispheres disconnected suggests that there are compensatory pathways which brain neural circuits develop in the callosal; this verifies the plastic nature of the nervous system, the "being and becoming" is always under operation and in many instances this "becoming" results in changes that compensate a dysfunction.

If any action initiated by any module (those circles in the figure) will be interpreted by the left hemisphere, then it probably contributes to the illusion of free will discussed in the preceding section. Not only that, in our everyday life we often try to find explanations for the behaviour of others and of ourselves. Although most of us are not split-brain individuals some readers will have experienced occasions when we assert that this or that event occurred or we offer an interpretation on some events that we think happened in the past, only to realise, a few hours or perhaps minutes later, either upon a slight mental deliberation or because we are told by some, that the events did not quite happen that way, that it was a construction of our minds based on some memories we had and the current circumstances. And yet, at the time we declared those things it seemed so real to us. While this example is not quite like those of split-brain cases, it may offer an illustration of a possible confabulation by our interpreter—perhaps the left hemisphere with all its language processing abilities—based on some memories stored in certain neural networks that become activated by the momentary conditions that induce us (that is, our conscious loquacious left-side interpreter) to declare something was like this and that. The tricks that memory plays on us may not be due to memory itself—that is, difficulty to retrieve or partial loss—but rather about what the interpreter in our brain decides.

One question that comes to mind is about the sense of self in split-brain people. How could the split-brain patients not experience any disruption in their experience as a unified self when the two hemispheres are physically and functionally disconnected? We have already said that in real-life situations they behave mostly like everybody else, because life does not occur in a scientific laboratory where these individuals can be specifically tested; for one thing, the integration of information can occur by behavioural mechanisms such as scanning eye movements that provide left and right (visual) unification in both hemispheres. Some of these mechanisms, non-verbal ones, are in fact used as tricks by some patients when they are tested, like that individual whose left hand was holding different geometrical shapes (without him seeing these) with the aim to verify whether or not his right brain side could say/identify what shapes those were, and as it happened the researcher noticed the person looking around the room when an object was placed in his left hand and, shortly after, he was able to identify the shape of the object, whether circular or square etc. It turned out the patient was cheating: he was looking around the room to identify with his left hemisphere the shapes in his left hand that were sensed by the right hemisphere, so when he handled a sphere he would look until he (his left brain) saw a clock on the wall and hence, was able to pronounce the word "circular shape"; his right brain knew the shape but could not verbalise it, nevertheless this right hemisphere could direct his eyes to help the left brain to identify it. Very smart right brain, isn't it? This cheating has even a technical name: cross-cueing (observations described in J. E. Bogen's *Brain Circuits and Functions of the Mind*). At the same time, another reason why these patients do not suffer a fragmentation of their personal identities is that even with a callosum totally sectioned there is the possibility of cross-integration of information by systems of the intact brainstem which can, among other things, help in the cross-integration of mood/emotional states, or as well through other subcortical commissures, not part of the callosal fibres, that were briefly mentioned at the beginning of this section. In conclusion, these patients should not present any striking alteration in their sense of personal identities, their interpreter still keeps things in order.

It may be of interest to mention what could be almost the opposite to the split brain, namely two people who share parts of their brains. We will not go into this phenomenon but interested readers can find the extraordinary case of Canadian twins who share a thalamus—technically known as conjoined craniopagus twins—on some websites[1]. As can be expected, their brains share information, for example, it has been reported how one can feel and taste what the other is experiencing. So here the idea of the self and consciousness as a private thing goes down the drain, and we can only hope that, notwithstanding some people's attitudes which this

[1] https://en.wikipedia.org/wiki/Krista_and_Tatiana_Hogan

unusual condition could originate, their life will be not only enjoyed but doubly enjoyed.

To finalise this section, some main conclusions that can be derived from a variety of split-brain studies and that bear on the issue of the brain self-model are the following: (1) each brain hemisphere simultaneously interprets the significance of (visual) events and reaches independent decisions about appropriate motor responses to those events; (2) the separated hemispheres can function independently and in parallel, and each has its own cognitive style; and (3) the normally dominant left hemisphere, perhaps thanks to its linguistic and semantic capabilities, interprets actions and events in efforts to maintain the unified perception of the personal identity and conscious awareness.

3.1 Closing the Coffin of Free Will

After what we have described in Chaps. 2 and 3, now it is time to proceed to add the final nail to that notion of free will as the ability to consciously perform unbiased choices and actions. It has become apparent that whereas we can make choices and decide on a course of action, these choices are determined, many times, by forces we cannot know because the brain's computations remain, in their majority, unconscious to us. We just cannot be aware of everything going on in our minds. As the cases of patients D.F. and J.W. showed, it seems like multiple decisions are made, each in its own brain region, and which one comes out, which "decision" is reported by the person, depends on how he/she is asked: the reportability issue mentioned in the preceding paragraphs. For the purpose of adding the final nail, we must mention one fact that bears on an issue that much of humanity is trying, and has been trying for a long time, to amend: that of racism or xenophobia, or prejudice.

It should be known and acknowledged that the biological roots of these dispositions are present in the genes and in the neural networks the genes "manufacture". So it happens that brains are attuned to one fundamental feature in those moods, skin colour. Neuroimaging experiments have shown that the amygdala—a deep, subcortical brain area involved in the processing of emotional responses including fear, anxiety and aggression—becomes more activated after flashing a face of a different race for just a fraction of a second, in fact, the presentation is so short that it is almost a subliminal stimulus. The widespread connections from the amygdala to the cortex and many other brain regions, causes this rapid detection of emotionally relevant stimuli to reach all those ventriloquists we previously met. The amygdala, then, is announcing to the rest of the brain that a potential danger, in this case, a very different individual, is in front of us. It sends the message to the body too via neural networks of the hypothalamus to which it is connected—the hypothalamus is a collection of nuclei responsible for the regulation of many processes in the body. Those interested in details about how brains perceive race can read the review by Kubota and colleagues "The neuroscience of race" (Kubota et al., 2012). While this neuroscientific observation may be discouraging, even depressing, to some, there are options to alter the prejudgments instilled by the amygdala and related brain regions, for we have a prefrontal cortex that, remember, is an association area that combines perceptions of different modalities into a unified construct and regulates the behaviour in a conscious manner. Thus, in principle, it could override the emotional charge imposed by the amygdala. In fact, more neuroscientific evidence comes to the rescue: if those images of faces are shown for a longer time, not at the "subliminal speed", then the activations include other brain areas besides the amygdala like, precisely, the prefrontal cortex, so in principle, this cortical lobe could exert an executive control of the amygdaloid response.

What does it all mean? Our brains, like it or not, have an innate ability to quickly detect entities that do not belong to our group or race, but that with some reflection we can abort some prejudices and actions that may be determined by that fast detection of a different person. Not that this is an easy task, for these preconceptions and biases are firmly established in our brain neural circuits. Albert Einstein, who was Jewish and renounced his German citizenship in 1933 due to Adolf Hitler's rise, is quoted as saying: "It is easier to break the atom than to destroy prejudice". In the aftermath of the first world war, Germany's economic devastation and social situation made it easy for Hitler to exploit the brain's quick ability to judge "the other" and blame the Jewish population, among others, with horrific outcomes that still resonate today. But it should be always possible to reason about our immediate feelings such that a more reasonable behaviour is exercised. This is the point that has been and will continue to be emphasised in this book, that thanks to the power of our intellects we can now go against the will of the genes and the "bad" ventriloquists the genes have produced. The use of the word "bad" here is in a social context (in nature nothing is good or bad, things just are). So if the bases of racism and xenophobia are in our brains, the solutions are also in our brains.

In any case, this example of rapid activation of certain brain regions upon perceiving some people, things or events and its consequences on almost reflex-like behaviours, adds more to the already immense evidence that free will as commonly understood is a fallacy. Moreover, our freedom in general—at a higher level of description, so to speak—is very limited because we are born in a society and in an environment. As our developing neural circuitry is in constant reciprocal interaction with our environment, to a large extent we are the products of our time. Nonetheless, it feels good to have that very little freedom. A book reviewing the subject and showing how our modes of thinking and situations may force our responses and behaviours is D. Ariely's *Predictably Irrational: The Hidden Forces That Shape Our Decisions* (Harper Collins, 2008).

For the sake of fairness, we have to mention that some ideas that have been here discussed are not products of new insights of the past few decades based on sophisticated neuroscientific experiments, but that have been around for quite some time. We have already quoted scholars like Hume or Locke with their views on these topics, and we will end with the words of B. Spinoza, considered one of the great rationalists of the seventeenth century, that resonate with much of what we have concluded. In a letter to a certain G. H. Schaller dated October 1674, he said: "*Further conceive, I beg, that a stone, while continuing in motion, should be capable of thinking and knowing that it is endeavouring, as far as it can, to continue to move. Such a stone, being conscious merely of its own endeavour, would believe itself to be completely free and would think that it continued in motion solely because of its own wish. This is that human freedom, which all boast that they possess, and which consists solely in the fact that men are conscious of their own desire but are ignorant of the causes whereby that desire has been determined. As this misconception is innate in all men, it is not easily conquered. For, although experience abundantly shows that men can do anything rather than check their desires, and that very often, when a prey to conflicting emotions, they see the better course and follow the worse, they yet believe themselves to be free*".

Indeed, and in spite of all we now know about the neuroscience of will and even if we accept that the common concept of free will rests already buried in a completely locked coffin, it is still so hard to live our lives without that notion. As Robert Sapolsky aptly comments in his book "I can't really imagine how to live your life as if there is no free will. It may never be possible to view ourselves as the sum of our biology" (Sapolsky, 2017). The belief in free will may be one more psychological need of ours, like the need to believe in an agency, in a self and in the meaning of life. We may have to apologise for our attempts at the demystification of these psychological needs, and while this book is not meant to instruct people how to live their lives and what to believe, one thing we do believe is that better to be informed and knowledgeable than ignorant and superstitious.

3.1.1 A Cry for Freedom

As a postscript after discussing where the freedom of will lies, a few words about freedom in general, may be worth. Humans, and animals too, cry for freedom. Freedom to act, to behave, to think. Yet, as much as the will has limited freedom, this freedom of behaviour is constrained too by many factors. The fact we live in society imposes many restrictions to our freedom. We cannot do things in our neighborhoods that we would do living alone in the mountains, but as well, even in the wilderness, our freedom would be limited by the flora and fauna around, as much as it occurs with any other animal. Complete unrestricted freedom is thus another illusion. Nonetheless, in most of our modern societies, we are relatively free. Still, it is interesting to see that some populations crave more freedom. A certain group living in a certain area of a country desires to become independent, and we have witnessed the calls for freedom of populations in, say, Quebec in Canada or Cataluña in Spain. And this is interesting because this desire for more freedom is another manifestation of that ancient, primaeval disposition that the genes carved in our brains. Our tribal nature still remains alive in spite of such a long history; a nature that has its roots in those primordial instincts of joining in groups of conspecifics.

Is this attitude still relevant in this time and age? In those two aforementioned illustrations of specific populations wishing more independence, one wonders what the significance of independence would be, what changes in life would exist for the average Catalan or Quebecois citizen in the street; obviously, changing the allegiance from one state to another, paying taxes to another government and few more things (perhaps the most important change for some is that they would have their own sports teams!). The reality is that not too many things would be different for those average citizens as they live their daily lives in the case their regions reached independence because, at least in most of the Western world, anybody can behave freely in any country, speaking their own dialects, eating their own food and drinking their own wines. All done with due respect for the others, of course. It may be different in other parts of the world, where the populations living together are widely different and, in many cases, violent against each other. This disposition for craving more independence was more relevant in ancient history where neighbouring tribes had constant conflicts—astute generals took advantage of these enmities to conquer vast areas, like Julius Caesar did with the Gallic tribes or the Europeans in Central and South America. This is not a political text to endorse one or another government, these aspects are mentioned here because they represent an ancient echo of a most primordial instinct, and it is remarkable that, in spite of our advanced intellects, some of us are still enslaved by these drives resulting, sometimes, in irrational and disastrous consequences. We tend to make frontiers rather than bonds, pushed by these ancestral drives that make us join conspecifics and avoid those that are different. Would this (intellectual) tide recede if we were to use our highly developed brain neocortex? Time will tell.

Let us keep sight of what got us to this point. So far, we have discussed how the self, our sense of personal identity, emerges as a natural consequence of our embodied brains immersed in an environment, and how this perception of unity in cognition and behaviour will develop and change as the child grows, and in shorter time scales as a consequence of the mutual interactions among various brain regions which process different types of information. All this is an exemplification of the mind's "being and becoming", the models created by our brains of the self and the surroundings constantly being updated. But these changes do not necessarily need months or years to occur, selves can change every day, every hour or even in a few seconds, which is what we will see in the following chapter where the astonishing experiences of patients with multiple personality disorder—today known as dissociative identity disorder—will be described to try to understand how within a few seconds their identities are totally transformed.

Chapter 4
Dissociative Identity Disorders. Are Multiple Personalities Ever Lonely, or Can a Psychiatrist Charge Twice a Person with Dual Personality?

The previous sections have enlightened us a bit to understand better who we are; the left-brain hemisphere, the right brain, the basal ganglia and amygdala and many other deep structures, the spinal cord and the sensory terminals, all weaving a model of the self, a sense of personal identity and with that a model of our surroundings, reality as we call it.

In this reality, though, we perceive ourselves as one, in spite of having different brain neuronal networks that work on different computations and various ventriloquists, and, as we saw before, the two halves of our brains have their own "personality"; but even when separated, the two halves still seem to function to create a unified identity. Is it possible that there are instances when this perceived unity is shattered? Well, yes and no. Learn about the amazing stories of people who have many selves between their ears.

Dissociative identity disorder (DID), previously known as multiple personality disorder, is characterised by the presence of two or more distinct identities or personality states that recurrently take control of the individual's behaviour. According to the technical description offered by the Diagnostic and Statistical Manual of Mental Disorders, "the essential feature of the dissociative disorders is a disruption in usually integrated functions of consciousness, memory, identity or perception of the environment. The disturbance may be sudden or gradual, transient or chronic". Dissociative disorders include dissociative identity disorder proper and as well other syndromes like depersonalization disorder (recurrent feelings of being detached from one's mental processes or body, the patient feeling like a zombie) and dissociative amnesia (the inability to recall an important piece of personal information).

One remarkable fact is that the identities or personality states, known as the "alters", differ enormously from one another, each having "its own relatively enduring pattern of perceiving, relating to and thinking about the environment and self" (in words from that Manual again). Dissociation here means a disruption of the

J. L. Perez Velazquez, V. Nenadovic, *Being and becoming*,
https://doi.org/10.1007/978-3-030-78264-1_4

usually integrated functions of consciousness, memory, identity or perception of the environment. The alter-personalities can be conceptualised as brain states, each organised around a prevailing effect, a sense of self (to the point of having a distinct body image), and a set of autobiographical memories. The switching between the alters can happen in seconds, but it can also take minutes or even days. These personalities naturally have different identities, they have their own age, gender, name, even race. Each has his/her own postures, gestures, ways of talking; sometimes the alters are animals. Different personalities can also have different medical conditions. As each personality reveals itself and controls the individuals' behaviour and thoughts, one has the impression that a completely different person from the one a few seconds before is in front of us; and in fact, it is true, for this is a brain with different neuroelectric patterns (see below about EEG recorded in these patients) and therefore it is a different self, different person.

The formation of DID "alters" is thought to involve the development of multiple centres of "information processing" in the mind (now it is not the time to discuss what is meant by neural information processing which has several technical aspects to it, so we are using it in the most intuitive sense). Each alter is capable of independent thoughts, emotions, sense of self, memories and distinct ways of behaving. The dissociative identities seem to be relatively stable with enduring patterns of behaviour that are largely unintegrated with each other and are often in a direct conflict. Indeed, some alters have a hostile relationship, which is not too surprising: different people—that is different "alters" each inhabiting one brain—tend to have different opinions and often hostility arises among them, so why not among the alters occupying the same brain. It should be understood that these alters are, most of the times, really different persons. Different persons that come even with different physiologies! As amazing as this may sound, it is true.

Measurable physiological differences between the personalities in DID have been consistently reported. These include differences in handedness, so one alter may be left-handed, the other right-handed; in visual acuity, oneself may be short-sighted, another may have a normal vision; in the sensitivity to various visual, tactile, olfactory and auditory stimuli. Can all this be true? Can a right-handed person suddenly become a skilled left-handed one? Could it be these patients are making believe all these things? Whereas some behaviours could be faked, others not. Thus, it is impossible to fake an allergic reaction: it has been reported that a patient will experience allergies during the manifestation of one of his/her alters. Different responses to the same medication have been observed too, such that (in simple words) when alter A is present the medication works fine, when alter B shows up the same medication fails or has side effects. This phenomenon has enormous implications for the origin and treatment of chronic illnesses and conditions, in that the potential of manipulating brain networks to treat illness and even correct vision without lenses or surgery would be a great advancement to humanity.

Also, there are case reports of distinct autonomic nervous system indices in DID identities, including galvanic skin responses, heart rate, blood pressure, etc. Even neurophysiological differences across altered personality states have also been reported, namely, EEG recordings being different when one alter is in charge and changing to another electrical pattern when another alter appears. Not only electrical activity in the brain, but also brain metabolic activity—normally measured using neuroimaging techniques like functional MRI, or fMRI for short—changes during the personality switch, variations that are more pronounced in certain brain areas like the hippocampus, suggesting that the personality switch may result from changes in hippocampal and temporal lobe function; this may be expected because the hippocampus is a region involved in memory and, as aforementioned, different alters will have different memories.

These (neuro)physiological alterations are all astounding. So that you understand what this all means, picture your body experiencing, say, an allergy to some compound showing red patches in your skin only when a certain personality, certain alter, is manifested, and the red spots going away when another alter, perhaps the original self, appears. There is no way you can fake that. That means this specific "allergic" alter is using some brain areas where there is communication between the nervous system and the immune system. For instance, the "allergic alter" commands the sympathetic and parasympathetic nerves that innervate the immune system causing the release of histamine, a chemical mediator, in response to the presence of the allergen. This network is not activated when the "nonallergic alter" is present. Organ systems in the body, just like the brain, are connected to everything else, such that you may have an upset stomach when you receive bad news, or you may feel happier when your stomach is full and your appetite satiated. Because each alter takes over the brain, therefore patients with DID report perplexing fluctuations in skills, habits or well-learned behaviours such as athletic abilities or fluency in a foreign language—one alter may know French, another Spanish, one may be good at tennis, another at football. As unbelievable as these things sound, they are true. These observations should make us realise the immense power within our brains; and, on second thought, perchance one does not need to suffer from DID to recognise and use this power. Unfortunately, we tend to cling to beliefs that impede us to achieve what seem to be miracles. More on this in Part III.

On this topic of beliefs, we take this opportunity to clarify, demystify, another popular myth, so let us pause the fascinating discussion of the manifestations of DID for a moment to further illustrate how our beliefs hold sway. This digression is also prompted by what we are seeing in these patients, in that their brains have an enormous capacity to hold many selves, in some cases the number of alters can be 100 (the average is 10). This indicates that their brain's capacity is being used to a great extent. A long-held false belief is that we use a small portion of our brains, 10% more specifically. This is not true. The brain is continually monitoring the state of the body, the incoming sensory stimuli, our movements and actions, and all these tasks imply that each cell of the brain is, basically, continuously active. Even when we sleep, the brain thalamocortical networks—a major part of the brain's anatomy—are involved in the intense activity as can be observed by performing intracerebral electrophysiological recordings. However, this intense synchronous neural activity during deep sleep (slow-wave sleep in technical terms) does not have the organisation to process adequately information to maintain conscious awareness, which is fine because we are sleeping. So rest assured that your brain never rests and the whole organ is almost constantly functioning. But whether this intense neural function results in the creation of works of art and symphonies, in the finding of amazing scientific discoveries, in the design of spacecrafts or in people becoming chess-masters prodigies, is another matter altogether. The fact that most of the brain is functioning does not guarantee that everybody becomes a genius or that we will never make silly mistakes.

The origins of this fallacy about using only a small fraction of the brain are several, starting with the reserve energy theory of W. James and his student B. Sidis in the late nineteenth century. But one more modern reason stems from the results presented in typical neurocognitive experiments—and perhaps as well from our eerie feeling that we do not think too much in-depth and make many mistakes! When scientists started to take brain scans during experimental conditions—around the 1970s—and show the figures coming from the analysis of those scans, the common figure (which anybody can see by perusing a typical cognitive neuroscience paper) is a brain mostly black with a few colourful spots. Technical note: the brain's activity in these studies is colour-coded, with black representing no activity and red maximal. The colourful spots represent where most of the neural activity takes place in the particular experiment being carried out in that study. Hence, someone looking at those pictures will think that only a tiny fraction of the brain is used. But what they ignore is that the pictures are generated after a sophisticated analysis where thresholds are imposed, many times rather arbitrarily, which will determine what activity is above the threshold, hence "active" and which one is not, or "inactive". Depending on the chosen threshold, your brain after one of these scans may appear all black (dead) or all red (big seizure). It is all that relative. Hopefully, the fallacy has been clarified, so beware of those who claim they can sell you a method to unlock your hidden brain powers beyond that famous 10%.

Why did we step away from our DID discussion for this interlude? Well, we didn't. Ironically, while even in people with DID their whole brain is continually active, what emerges at times is a "fraction" of what would have been an integrated self.

How do the alters switch? Someone may think that having distinct personalities may be advantageous. Think of an original personality that is timid and fainthearted, well then, it could be a good thing to bring about another strong and assertive alter when the situation arises, such as when confronted with a bully. It is not that easy, and in general, this syndrome is tremendously excruciating for patients, the only advantage the various personalities offer could be the escape from a reality they do not accept, as explained below when discussing the causes of this condition. During examination and therapy, the alter-personalities are drawn out by the psychiatrist, who will determine the diagnosis by observing the number of alters present in the patient. It is sometimes truly amazing to witness how a well-trained psychiatrist pulls one alter after another. In normal everyday life, alters will materialise depending on a variety of factors like stress, looking at photographs, recalling memories, etc. Some patients are able to recognise the signs of switching. And in the same manner as alters are not created by conscious choice or planning, normally, the switching does not occur by choice. How does a patient notice? The patient, meaning the original self (admittedly at this point, after the split-brain section and this one, it is becoming hard to keep track of who, or what "ventriloquist", we are talking about or addressing!), will be aware of symptoms that typically include mannerisms present with different alters, severe memory loss (because what one alter does may not be available for future recollection by the other alters and the original self), different vocal expressions, mood swings, flashbacks of trauma and others.

Now we offer one example of the experience of meeting one of these patients. There is a story of a gentleman who frequented a park with his dog, a large husky. He got to know other dog owners who would bring their canine companions for exercise. He confided his diagnosis of DID to one woman, who then understood the seemingly different "men" that she would meet accompanying the dog. Of interest, during one emergence of an alternate, she did not initially notice a change in the man, but his dog ran up to him and began barking; the dog continued barking and running back and forth in front of his owner. Could it be that what was imperceptible to the woman was clearly evident to his dog? The man turned to the woman next to him asking her who she was and what year it was. He then began yelling at her to get away from him. He did not return to the park for several weeks and when he did, he greeted his acquaintance with no memory of the previous event or his absence from the park.

Very little is known about brain structure and function involved in DID, particularly with regards to mechanisms of personality switches. DID is one of the least understood mental syndromes, and due to its extreme presentation has been the subject of many sensationalised movies that reinforce misinformation. But about the causes for this illness, those are better characterised. Hence, the dissociative aspect is thought to be a coping mechanism; the person literally dissociates him/herself from a situation or experience that is too violent, traumatic or painful to assimilate with their conscious, original self. DID has been strongly linked to severe childhood trauma, such as physical and sexual abuse (as reported in about 85% of cases). The essential hypothesis linking trauma and dissociative symptoms is the failure of the integrative capacity of the brain, resulting in abnormal memory processing with the inability to integrate and synthesise emotions and sensations related to traumatic experiences into a whole. Since the syndrome is normally based on early traumatic experiences, many symptoms appear early in childhood. It is thought that the child learns to use dissociation in order to cope with severe stress, using it as a sort of "psychological flight" reaction.

Conceivably, as it occurs in epilepsy where seizures are just a manifestation of an abnormal brain that is epileptic (or likewise fever is a consequence of microbial infections), having separate identities could be merely the side effect or consequence of something greater, not the exclusive disorder. The real dysfunction lies in the trauma and the effects it had on the child's mind because as it is constantly stressed in this book (sorry again to belabour the point, but it is fundamental to comprehend self-awareness), brains are malleable—they are and they become—and their neuronal connectivity and function changes depending on behaviours and the events in the surroundings.

It then should come as no surprise to find out that DID occurs in other neurological syndromes where there is aberrant neural activity. As such, it has been seen in epileptic patients. A classic paper studied 12 cases with EEG and clinical manifestations of epilepsy, and the author describes each case in detail with explanations about their alters' personalities, hence it is a recommended reading for those whose interest in the DID condition has been aroused by our text; the title of that paper is "Dissociative states with abnormal temporal lobe EEG—Multiple personalities and the illusion of possession" (Mesulam, 1981). Now, a great deal is known about brain activity in epilepsy but because not all epileptics have DID we cannot draw too many conclusions about the neural mechanisms of DID based on the known facts about seizures. Perhaps one thing should stand out, that of disorganised neural activity during epileptiform events, so we are talking again about a lack of coordinated neural dynamics present in the epileptic brain that may result in various symptoms like seizures and alter personalities.

Dissociative disorders like depersonalization can also be induced by drugs. For example, ketamine, cannabinoids, opioids, hallucinogens like LSD and psilocybin can have dissociation as a side effect. Chronic depersonalization has also been associated with tumours or migraines, and of course, the aforementioned epilepsy.

Having reviewed in the previous section the split-brain phenomenon, one may think that there is a parallel between that condition and the multiple personality syndrome examined here, after all the split-brains seem to have two almost independent brain hemispheres doing their own computations. In reality, these are two very different situations. In DID we have seen that each alter remains fully aware of their actions and thoughts, they all possess self-awareness. In contrast, in the split-brain patient, there is only oneself that interprets events and actions—which may be performed consciously or unconsciously—and then there are parts of the brain that perform computations that remain out of the awareness of that interpreter. In fact, as a thought experiment, it would be impossible to create a DID patient by sectioning and separating areas of the brain because the various alter-personalities all draw from the same brain neurophysiological resources and must have access to areas where memories are kept or auditory information is processed or where the motor cortex sends the final signals to move the muscles.

If the general underlying neural mechanism (the dysfunction) in DID is a loss of integration of various brain areas that may be needed for a cohesive conscious experience, then the question arises as to whether there can be a remedy to restore the normal integration. Can DID be cured? It is clear that the several alters within a brain must draw from the same biological resources. After all, there is only one brain, and thus there should be one core self, in spite of having various selves each one feeding on some specific neural networks where memories are stored, that being the reason for the different autobiographical alters. The path to healing then could rely on gaining access to all that information, as well as establishing meaningful communication among the alters. In fact, sometimes patients learn to cope and the various personalities integrate into one, although, again, the neurophysiological mechanisms are unknown. Therapeutic practices include having the patient recognise and accept the existence of the alter-personalities, coordinate these selves in some sort of psychological integration, perhaps try to get rid of some alters that create negative influences, or change behavioural habits because it is possible that some alters may then lose their reason for existing. Whereas there is no real cure for the syndrome (at least at the time of this writing), some individuals are greatly helped and go on living almost normal lives thanks to psychotherapy. Does the success with the therapy in some of these patients indicate it is possible to abolish a self? Not so fast. Before jumping to conclusions, let us examine how tremendously resilient is the perception of the personal identity by considering conditions that, in principle, should erase almost all of that sense of unity.

Chapter 5
The Enduring Self, or How to Annihilate the Self

In the preceding section, a multiplication of selves in DID sufferers was examined, and as well, the success of psychotherapy with some patients was noted. Does this success mean the self can be destroyed, or annihilated? After all, the improvement in some patients is concomitant with a decrease in the number of alters, meaning some of these alter-selves have been obliterated. Yet, there are some, or at least one, that always remain(s) housed in the brain of the patient. One can think of some special conditions under which the perception of unity in cognition and behaviour that constitutes the self could make it very difficult for the brain to maintain the self-model. How about, for instance, memory loss, or amnesia? Can it be that problems with memory perturb the sense of personal identity? The self-model requires keeping track of what we just did or did in the distant past (as indicated in the previous accounts about mental aspects like an agency). It is because we remember what we did in our youth that we believe we are the same person as that teenager; hence, if memory retrieval is not possible, one could suspect the sense of the personal identity may be distorted to some extent.

Let us inspect cases of very severe amnesia, individuals who basically are living in the present. One such case is that of the British musicologist Clive Wearing (born in 1938). After suffering from encephalitis, his hippocampi were damaged. The two hippocampi, one in each brain hemisphere, are fundamental structures to create new memories, but old memories that have already been stored in some neural networks of the neocortex remain accessible to conscious recall. This is one reason why elderly people have problems remembering what they did yesterday—their hippocampi start to become dysfunctional and cannot establish new memories—but since these areas are not required to recall already formed memories, they can perfectly remember events from their distant past. Mr. Wearing had not only very severe retrograde amnesia due to the damage to the hippocampus but also anterograde amnesia perhaps due to injury to other cortical neural areas where old memories were

J. L. Perez Velazquez, V. Nenadovic, *Being and becoming*,
https://doi.org/10.1007/978-3-030-78264-1_5

stored. Anterograde amnesia is a loss of the ability to create new memories after the event/injury that caused the amnesia, leading to a partial or complete inability to recall the recent past, while long-term memories from before the event remain intact. Retrograde amnesia refers to the inability to recall events that occurred before the injury that caused the amnesia. As a result of these two forms of amnesia, Wearing's memory only lasts for about 30 seconds; therefore, he is very close to living in a constant present. The story has been written by his wife in *Forever Today* (D. Wearing, Random House Press, 2005), revealing how very different the life of her husband is as compared to normal people with unlimited access to memories. Despite having retrograde as well as anterograde amnesia, and thus only a moment-to-moment consciousness, he still recalls how to play the piano, and this is because his cerebellum, which is involved in procedural memory—and also responsible for coordinating voluntary movements as mentioned in Sect. 2.2.1—did not incur any damage, but as soon as he stops the music, Wearing forgets that he just played the piano.

One can appreciate better what it is to live in this situation by looking at the records in his diary, like in this example where we see that earlier entries are usually crossed out, since only a few minutes after he writes them he forgets that he did and dismisses them:

~~8:31 AM: Now I am really, completely awake.~~
~~9:06 AM: Now I am perfectly, overwhelmingly awake.~~
9:34 AM: Now I am superlatively, actually awake.

He is able to know certain facts about himself: he knows that he is a musician, that he has a wife (may not remember her, though), and he realises too he has a profound impairment. In spite of living basically in the present moment, accounts from his wife and others indicate that he loses the sense of the self only very momentarily, and then, it reappears, only to vanish again, but in general, it can be said that he possesses self-knowledge. Indeed, if we were to meet him and briefly talk to him, we probably would not notice anything peculiar other than he has little memory, but he would not impress us as someone with no self-awareness.

Another case where the sense of personal identity was specifically investigated is that of a person who suffered head trauma and has accurate memories of scenes from his past, but for whom the memories lack the sense of "mineness", so to speak. He is capable of episodic recollection but without a sense of personal ownership of those memories (Klein & Nichols, 2012). And like in Wearing's case, the sense of self remains, in his own words "Yes, I definitely have no identity problem [...]. I have normal control over remembering facts and scenes from my past. But when I remember scenes from before the injury, they do not feel as if they happened to me, though intellectually, I know that they did. They felt as if they happened to someone else". Here, then, looks like self-reference remained in the recollection, but the insertion of the self as an agent of the memory fails. Perhaps he had a problem with the identification of agency.

A study where several amnesic patients were examined and compared to healthy persons revealed that while the amnesic participants generated fewer personal semantic memories than the healthy individuals to support their self-statements, they still possessed a sense of self (Grilli & Verfaellie, 2015). And the most famous of all memory-impaired patients, Henry Gustav Molaison, better known as HM in the neuroscientific world, still had a sense of personal identity in spite of suffering anterograde and retrograde amnesia after brain surgery. HM has been one of the most widely studied patients from the late 1950s until his death in 2008. These studies have played a very important role in the development of theories that explain the link between brain function and memory and in the development of cognitive neuropsychology. Several areas of neuroscience owe their advancement to him, and today his brain, now sliced into preserved histological sections, resides at UC San Diego.

So, we already see that the self-model, or a sort of it, remains even in persons who live basically in the present. But at least one case study (case studies are normally short medical papers that evaluate only one patient) indicated that the patient suffered a persistent loss of personal identity (Piolino et al., 2005). This patient could be the exception that confirms the rule that the perception of the personal identity is extremely hard to annihilate. In this and in the past chapter, we thus have seen how resilient the sense of identity is; even in abnormal brains like DID sufferers, we find not only one but several selves; and this numerosity indicates the natural tendency of brains to create self-models. It is not that DID patients do not perceive a unity in behaviour and cognition (a self); it is that they perceive several "unities". The multiplicity of alters illustrates this tremendous tendency of brains to create selves. It is in this sense that our answer to the question posed at the start of Chap. 4 about whether there could be instances when the perceived unity of a self is shattered was "yes and no": yes because for sure, there is more than one sense of unity, so it can be said this unity is fragmented in the brain of the DID patient, there may be a feeling of being split up in various selves, but still, each of those preserves that perception of unity unique to each one. Hence, the perception of unity is not destroyed.

Then, does the self dissolves at some point? Well, it disappears every day, in fact, when we sleep—this occurs only in slow-wave sleep, the periods with no dreams. Remember that during rapid eye movement (REM) sleep, dreams arise, and then we are aware of ourselves and are perceiving an internal world created by our imagination. Self-awareness also vanishes during generalised epileptic seizures, when the disorganised neural activity does not allow the brain to sense and respond to the environment. And of course, the self is gone during coma and deep anaesthesia. Putting all these things together, one could conclude that the self fades away only when there are no perceptions: during anaesthesia, deep sleep, coma and generalised seizures, there is not much sensory processing, thus no perceptions. As asserted in Chap. 1, perception is everything, and the sense of self requires it too; thus, our proposed description of the self as the perception of a unity in cognition and behaviour makes sense in light of this evidence. A caveat: please do not take definitions too seriously because, as we will see below, strict definitions of these complex phenomena like consciousness or life are nearly impossible to formulate and tend to lead to misunderstanding more than understanding.

But let us go one step further; why perceptions are absent during coma, deep sleep and ictal events? Because the neuronal activity at those times does not allow the brain neural networks to process information appropriately. The specific description of these neuronal mechanisms is beyond the scope of this text; let us just say that the brain activity during slow-wave sleep, coma and seizures has some features that do not allow brain circuits to organise their connections in a fruitful manner for sensorimotor processing. Then, what is it? Is it lack of perception, or is it the neural activity that abolishes the self? It is both, since these are two sides of the same coin: neural activity determines perception, and perception determines neural activity (as we saw in Figs. 1.2 and 1.3 and in many other parts of this book). Therefore, this is a conundrum only for those who insist on strictly separating phenomena. In the moment one adopts a more holistic perspective and perceives the relations among things, these types of mysteries start to vanish.

There are conditions, though, when selves are on their way to extinction. A living experiment to witness the shrinking self is observing how dementia develops in someone who you know. Being with a person experiencing dementia is watching the progressive dissolution of the networks from which the person that one has known emerges. There is the loss of memory of others and past events. There will be neural networks triggered by sounds or images, and yet, this will be an isolated fraction of the original self, as can be inspected in a fascinating YouTube video showing a former ballerina with Alzheimer's disease in a wheelchair being triggered—so to speak—by music from Tchaikovsky's *Swan Lake* score[1]. What is captivating is that for a few moments, she executes not only the arm and upper body movements but also the facial expressions; thus, one can see how some parts of her brain still contain memories and how these can emerge if proper input—in this case, the music she knows—is given. During the progression of dementia, there can come a time when the individuals no longer recognize "themselves" in the mirror or in pictures. Furthermore, for some who had pre-existing conditions such as alcoholism, they can lose this part of their self; their identity of that of an addict is gone.

In any event, if it were possible to do it during normal daily activity, should we try to erase our sense of self? After all, there are (evolutionary) advantages of the selves. Harking back to Sect. 2.1, some benefits of having brains (and selves) were noted, in which these emerged from the primordial necessity to situate the organism in the environment, and developed through the perceptions of our behaviours within that environment. Realising our personal identities confers an adaptive value because we can situate ourselves, or any other animal with self-awareness, in space and, for those entities with more evolved brains, situate as well in time, which aids the organism to find prey and avoid becoming prey. In Sect. 2.2.1, other benefits were remarked. Plus, naturally, there are associated gains because we can proceed to self-evaluation or self-monitoring thoughts that may help us enjoy our lives. Indeed, one reason everybody searches after the personal identity is that it is a psychological need, as much as finding a reason for our lives. But recall from the first

[1] Ballerina with Alzheimer's hears swan lake, begins to dance - YouTube

paragraphs of this book the warning about the perils of the belief and subsequent strong attachment to a self, the emergence of the ego with the associated solid attachments to material things and people, and the psychological vulnerability this causes. Hence, while abolishing the self not only is nearly impossible but also may not be desirable, decreasing the size of the self/ego has profits too in order to more fully enjoy life, themes that will be more developed in part III.

Having clarified this point about the importance of perception for the sense of personal identity, a point which aids to demystify in part this business of the self, and before we proceed to demystify its counterpart, consciousness, let us take up one issue that has been floating around in all of this Part I.

Chapter 6
A Recipe for High Cognition: Are Consciousness and Self-awareness a Matter of Numbers?

In past sections we have introduced some properties of nervous systems, and especially brains that allow for the emergence of cognition and the related consciousness and self-awareness. An idea that has been permeating this Part I is that as more complexity—in simple terms more brain cells and connected neural circuits—develops in the nervous system, the more likely it is that high intelligence and the sense of personal identity arises. In Chap. 1 we examined the behaviours of primitive organisms with tiny nervous systems and those of more advanced animals that pass the mirror test. So, for a worm to behave and crawl around its world, it does not need a very complex system of mutually connected neural networks, a reflex-like neural activity suffices. We explained how the apparent self-awareness of the worm, recognising self-generated versus externally-generated stimuli, results from the phenomenon of corollary discharge (Sect. 1.5), for which a sense of self and agency is not necessary for the animal to recognise and avoid obstacles in its path or to retract and stay quiet because of a possible danger. And in Sect. 1.7 it was acknowledged that whereas we can endow these creatures some cognitive faculties as they possess some features of consciousness (like sensing and reacting to stimuli), it is only in the case of humans and probably other animals that the many nerve cells and networks with the appropriate connectivity in their brains allow them to go beyond the pure reflex, to display self-awareness and to have all the properties one can think of when enumerating the features of consciousness (some features will be detailed in Chap. 7 below when discussing why it is nearly unfeasible to define consciousness).

Can we then conceive that as more cells and networks are added to the primitive nervous systems, the characteristics of consciousness emerge in a graded fashion? Could it be that simple, just a question of numbers? In reality, adding more cells—neurons and glial cells (yes, please, do not forget in your cocktail the glial cells,

J. L. Perez Velazquez, V. Nenadovic, *Being and becoming*,
https://doi.org/10.1007/978-3-030-78264-1_6

recall their importance briefly mentioned in Sect. 1.4)—and making them work properly is not simple at all. It is not the absolute number of cells the important factor, it is the ways in which those neurons can communicate, the number and ways of connections among neuronal networks, connections that allow those networks to work together in an organised fashion. One can think about this in terms of a group of people trying to achieve something, adding thousands of individuals but without any coordination in their interactions—like everybody talking at the same time—will not be too helpful. It would be more efficient to create groups that work in coordination.

Is then a big brain smarter than a small one? Neuroscientific observations indicate that brain features that may explain the differences in cognitive powers amongst species are not based on absolute or relative brain size (relative to the rest of body weight) or a total number of neurons—humans have slightly more cortical neurons than cetaceans and elephants. And the same can be said about the number of synapses (as explained in Sect. 1.4, the synapse is a contact between two neurons), as elephants have about the same numbers of synapses in their brains as we do. What human brains seem to excel in is their neuronal packing density and, perhaps as a consequence, a bit larger number of synaptic contacts than other smart species. If you want numbers, here is a sample: there are about 0.9×10^{14} synapses in the brain cortex of dolphins and about 1.5×10^{14} synapses in the human cortex, a very close contest indeed. But do not lose sight of the fact that these numbers are very hard to determine with accuracy, and in general this matter of associating intelligence with brain features has always been a highly debated topic. Perhaps it is not a well-posed question since intelligence, for starters, is defined anthropocentrically as human intelligence (recall comments in Sect. 1.7). Hence, to summarise what could be a very long story, let us conclude that possibly the best fit between brain characteristics and degree of intelligence among mammals is reached by a combination of the number of neurons in the cerebral cortex, the neuronal packing density and the number of contacts—the synapses. Here we are again simplifying because there are other types of interactions among neurons that do not require chemical synapses, rather these interactions are due to the electromagnetic fields inherent in the neural activity. Bear in mind that what we are describing throughout this book, while true, is just an approximation. But science is nothing more than an approximation to the truth.

The purpose of this somewhat long digression is to explain that intelligence and high cognition emerge not so much because of absolute numbers of neurons but rather due to the connections among neurons and neural nets. Along these lines, some evidence obtained in the analysis of human brain electrical activity in several states of consciousness suggests that optimal brain information processing—that is, when people are fully awake and paying attention to the environment—is associated with a larger number of ways in which the neuronal brain regions can interact (Perez Velazquez et al., 2019). Again, we can easily understand this by means of the grouping of people simile used above: all people talking at the same time is not optimal and nobody talking is equally bad. The best-case scenario is when a coordinated communication occurs which is given by a large number of individuals sharing information. In the brain, as expounded in Sect. 1.4, it is all about coordinating the activity of the neurons, and this is why there is an important field of research that deals with the study of brain coordination dynamics.

We have thus seen that the brain is a densely interconnected mesh of neuronal networks that must organise, or coordinate, their activity within local circuits and across many brain regions in order to serve the purpose of performing adaptive behaviour. The capacity to generate adaptive solutions when faced with a varying environment is dependent upon the ways in which neural networks in diverse brain regions can interact. Our behaviours require processing a multitude of sensory information and combine it with information existing in the brain acquired from prior experience, processing that must be done with attention, reaching decisions, acting, and learning from the results of those actions. All of these computations require flexible and dynamic interactions across various brain areas. Therefore, to underline it again, it is not the number of neurons but rather the multiplicity of possible ways to connect to other neurons that is the key to higher intellect. And in case you wonder, it is the same story in the case of genes. Homo Sapiens possesses, according to the Human Genome Project, about 20,000 to 25,000 genes, and tiny worms have slightly fewer, around 20,000 genes; and yet, the human body is much more complex than that of the worm, so it seems that just as occurred in the case of neurons making up brains to show high intellect, the key is not the number of genes but the combinations that their activity can produce. For those readers unfamiliar with genetics: normally the product of a gene is a protein that will have various actions in the body, sometimes this action is about controlling other genes' products, to put it simply. In the words of the palaeontologist, evolutionary biologist, and historian of science Stephen J. Gould: "The key to complexity is not more genes, but more combinations and interactions generated by fewer units of code" (S. J. Gould in The New York Times, February 21st, 2001). Incidentally, all these facts about nervous systems and about genes mark one of the failures of the so-called reductionism in science, the collapse of the doctrine of one gene for one protein and one direction of causal flow from the genetic code to the totality, and indicate that it is very useful to consider not only the constituent parts—genes or cells—but also the nature of their interactions. Reductionism, very popular among scientists, rests on the idea of reducing complex interactions and entities to the sum of their constituent parts, such that the total is equal to the sum of the parts. This approach has its uses, but it is not quite practical when dealing with complex systems.

It would be unfair now to forget another aspect that promoted intelligence because, as many times emphasised in this book, neural activity and behaviours form a loop where one changes the other. Hence, a behavioural aspect that helped create good connected neuronal networks and thus better cognition is the bipedal posture or the liberation of the hands. As the hands were more free and able to act, their activity favoured the development of brain regions due to what is known as neural plasticity, a fundamental property of neurons that, in simple words, allow them to make more useful connections as their activity increases. It is the same with muscles: you work out and physically exercise hence your muscles develop, so if you intellectually exercise your brain changes too; and just having your limbs performing actions is enough for the neuronal networks to further develop their connectivity. The amazing consequences of this neuroplasticity, or the enhancement of brain areas to perform specific actions, can be witnessed in a large variety of phenomena, from the skills of piano players to the rehabilitation after stroke where brain regions which remained healthy are able to take over functions originally carried out by other brain areas that were destroyed due to the injury. And notice the loop here, hands performing actions improve neural connectivity that in turn improves those actions made by the hand muscles, exactly as, apparently, computer scientists Bo Dahlbom and Lars-Erik Janlert said: "Just as you cannot do very much carpentry with your bare hands, there is not much thinking you can do with your bare brain". All these considerations are the reason why the term "embodied brain" has appeared, and will continue to appear, many times in this text: cognition and all its consequences like consciousness, intelligence or self-awareness require a *nervous system* housed in a *body* immersed in an *environment.* If you remove one of these three items, there is not much cognition, consciousness or intelligence possible. The specialised, scientific name for the concept that organisms create their own experience through their actions is "enaction": we, and the rest of organisms, are not passive receivers of sensory information from the environment, but are actors in that environment, in the theatre of existence. If you want to see a sample of enaction you can watch this video[1]and learn how to ride the "Backwards Brain Bicycle", and at the same time, you will also learn that this learning requires not the logical understanding of how this weird bike works, rather entails a bodily performance. The intertwined nature of natural phenomena can never be underestimated, this is something fundamental to be realised if we are to understand nature, including our main topic in this first part of the book, the nature of the personal identity.

The anthropologist and polymath Robert M. Sapolsky said about the genes that "genes don't make sense outside the context of the environment" (Sapolsky, 2017); it is absolutely the same about the brain, its activity and function do not make sense without the context of the environment in which that brain is situated. It is in this sense that we can conceive that consciousness is a global property, not of just one brain but rather of the whole ecosystem, so to speak (more on this later, Fig. 7.3 and Chap. 7). Consequently, loops like the mentioned above about hands improving the brain and in turn the brain improving manual dexterity is just one illustration of this intertwined nature of phenomena. Loops are everywhere, we will see the (physical) reason for the ubiquity of loops in more detail in part II. And of course, loops are part of the emergence and creation of the self, the personal identity: we behave, perceive that behaviour, reflect on it which further makes us perform more actions the results of which we continue to perceive, and elaborate on these perceptions (and many times over-elaborate, in part III we will see why this over-elaboration of

[1] www.youtube.com/watch?v=MFzDaBzBlL0

thoughts/perceptions may lead to trouble), such that this never-ending loop proceeds while us barely realising it. And due to the recurrent connectivity among neural networks (the importance of this was described in Sect. 1.4), some neural loops have a self-referential character, which makes them interesting because if loops involve strong self-reference this may sometimes lead to very intriguing paradoxes; readers can learn some aspects of these strange loops in Douglas Hofstadter's book *Gödel, Escher, Bach—an Eternal Golden Braid.*

Some neural loops are depicted in Fig. 6.1, summarising very crudely the ideas or proposals that got us to this point about how one can go from the cognition and the behaviours of a worm to our elaborate behaviour and intelligence by, as it were, inserting a mind in the reflex. This self-referentiality at the anatomical level has a counterpart at the psychological level, so the following scenario emerges: if a brain captures percepts through sensory organs and is equipped with complex, recurrent, self-referential neural loops that evaluate these perceptions and elaborate on them, then the most typical of self-referential cogitations—the sense of personal identity—will naturally emerge. In fact, self-referentiality has been considered the key to consciousness, from Immanuel Kant in his 1781 book *Critique of Pure Reason*

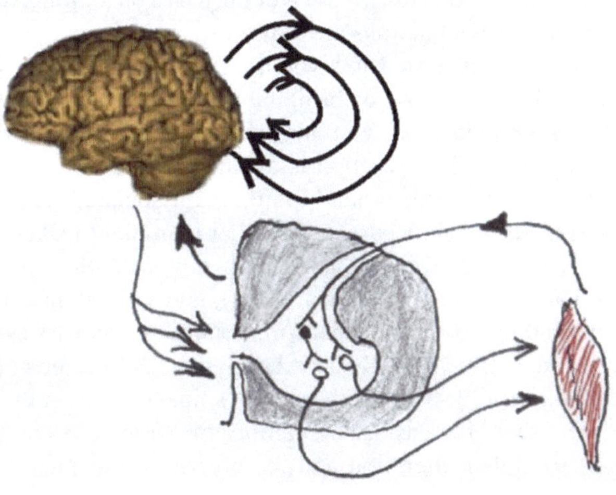

Fig. 6.1 Neural loops: a spinal reflex arc and the addition of brainy material. The typical neural pathway of the reflex arc from the muscle (in red) to the spinal cord (in grey) contributes to some of our actions without much thinking, like withdrawing the hand when it feels the heat. Connecting a brain to the spinal cord, with all its mutually interconnected neural networks, thus making more loops depicted by the arrows, contributes to adding some cogitation to our actions such that in principle we can consider whether or not to withdraw the hand (but if the heat intensity is very high then the fast neuronal activity in the neural chain of the spinal reflex arc will bypass the brain, thankfully, because thinking is slow and sometimes deleterious to the organism living in a fast world). As explained in Sect. 1.4, the recurrent anatomical connections among neurons create loops everywhere in the nervous system, and these structural rings can give rise to self-referential thoughts at the psychological level, such that the structural/anatomical and the psychological levels are closely related. In current neuroscience, it is thought that consciousness results from the distributed and reciprocal neuronal interactions across the brain

with his concept of apperception—the awareness of the process of cognition—to William James in his 1890 *Principles of Psychology*, and to the Buddhist sixth sense, the thinking about what is perceived via the other five senses.

The reflex arc achieved in the connection between the spinal cord and the muscle portrayed in Fig. 6.1 represents what is called implicit behavioural responses, that is, involuntary reaction to stimulus. If the neural activity (the information) reaches the brain and it has the time to evaluate the information and decide on a course of action then we talk about explicit behavioural responses. This spinal reflex arc explains how after spinal injury, a man may be capable of achieving an erection, even while the more complex sexual response—for which a brain is needed—becomes altered. Therefore, as explained in previous sections, the addition of the abundant neural circuitries in brain tissue allows the organism a finer regulation of overt responses (recall Sperry's quote in Sect. 1.2). But please note that the fact we have a brain attached to a spinal cord does not preclude some of our actions or thoughts still having a reflex-like character, as also explained in previous sections and especially in 3.1 where it was noted that whereas automatic stereotypes and prejudices are amenable to voluntary control, in reality many of these preconceptions are so firmly established in our brain neural circuits that materialise without much cogitation in a sort of reflex activity. Not only preconceptions or higher-order thoughts, but also more trivial actions are caused by reflexes in our brains. We can illustrate this and at the same time show you how your brain knows something you (consciously) probably don't know, by bringing your attention to what you do when you smell or bite into rancid food; you narrow your eyes, wrinkle your nose, and may vomit or have nausea. This reaction is dictated by the brain region called the insula (in Sect. 1.8.1 we met this crucial brain area associated with the subjective awareness of the self) that, as it occurs in any other mammal too, becomes immediately active when you bite rotten food and instructs your motor cortex to perform those actions so that your eyes, your nasal cavity, and your stomach are protected from the possible pathogens in the rancid material. All this bypassing your conscious awareness at the start, and only a bit later you will become conscious of the bad food. So, thank your reflexes, whether in the spinal cord or in the brain!

As we have seen, then, increasing the number of nerve cells and the ways they can interact leads to higher cognition and the emergence of most of the features associated with consciousness, one being self-awareness. So, where can we place the self in Fig. 6.1? Discussion of the brain areas involved in self-processing may have raised a question in some readers about whether there is an essential brain area in charge of self-awareness. For those with a vast thirst for knowledge, we will say that there is abundant literature on the topic, for instance, a good start could be Northoff & Bermpohl, 2004, where the authors propose that some neural networks located in midline structures of the brain are crucial for the processing of self-referential stimuli. We have also seen in this text how important some cortical regions are—like the parietal and prefrontal lobe association areas that integrate incoming sensory information—in terms of processing information related to the sense of agency, our place in the environment, and initiating voluntary actions; as well, internal, subcortical structures like the basal ganglia have their say in the perception of the self. If one continues to gather papers on this topic and marks the brain areas mentioned in those studies, one will end up with basically the entire brain marked, including even the cerebellum. And not only that, even beyond the brain, the spinal cord is fundamental to receive sensory information and transmit motor actions to the muscles, as we just saw in Fig. 6.1. So those with a holistic inclination will be happy to know that essentially all the nervous system is involved in self-awareness, and those who prefer a more specific and narrow perspective will be content as well to know there are some brain areas perhaps more directly involved in this feature. Hence, this is the picture that emerges at least at the time of this writing, perhaps next decade there may be a major discovery showing that a region in the central nervous system is crucially involved in the process, but for now, we can conclude that if you want a healthy sense of personal identity, be sure to take care of all your nervous system. And always remember that this is just a perception, a changing experience that should be taken with a grain or two of salt.

Now that we have examined the emergence of selfhood and have naturalised, and therefore demystified to some extent, concepts like free will and the personal identity, to end this Part I let us proceed to demystify consciousness.

Chapter 7
Demystifying Consciousness

Contents

The title of this section could have been the subtitle of this book, but, as there is virtually nothing new under the sun, it already exists (well, almost): "*Consciousness Demystified*", by Todd Feinberg and Jon Mallatt (MIT Press, 2018). Not too long ago, consciousness research was considered an esoteric pursuit. Today, the situation has dramatically changed as this area of research reached a few years ago an aura of scientific respectability and these days one may find many international meetings devoted to this topic and several scientific journals that accept without reservations papers on consciousness. Thus, if you were to attend one of the international consciousness meetings, you would find an assorted crowd of neuroscientists, physicists, biologists, engineers, neurologists, anesthesiologists, philosophers and perchance a few characters from other trades apparently totally unrelated to the topic like lawyers and mathematicians. Welcome to the multi-ring circus of consciousness research.

Naturally, consciousness and self-awareness are intimately related, the latter being one feature, or element, of the former. Some readers will have probably noticed some words that tend to accompany texts on consciousness, the most prominent perhaps is 'mystery'. This word, being also a common qualifying when talking about the brain, comprises several attitudes found in many of those characters attending meetings aforementioned in the previous paragraph. The physicist M. Gell-Mann pertinently said, "Consciousness is often seen as requiring a special kind of explanation, but the various aspects of self-awareness can presumably

J. L. Perez Velazquez, V. Nenadovic, *Being and becoming*,
https://doi.org/10.1007/978-3-030-78264-1_7

emerge when certain levels of complexity are reached in an organism: it is not necessary to assume additional mechanisms of hidden causes" (Gell-Mann, 2001). Because this book endeavours to naturalise all these "mysterious" phenomena like self-awareness and consciousness, and in part III death will be naturalised too, it will be our job in this section to explain what the mysterious part is in that notion we humans created long ago and called consciousness. For, make no mistake, consciousness, just like life and mind, are arbitrary concepts we have developed in efforts to encapsulate an immense variety of phenomena, a variety that may not be that easily encapsulated. Body and brain are organs, hence easier to define or describe; but how about life, or consciousness, can these concepts be defined in one or two sentences?

Part of that feeling of mystery stems from the attempts at defining consciousness in a short sentence with which everybody can agree. Let us start with considering this business of defining consciousness in one strict, clear-cut statement and what happens if, God forbid, we renounce to pursue such a precise definition. One should note that today there is no definition of consciousness that satisfies everybody, there is no single description accepted by the majority of scientists and philosophers of the mind. Considering that the notion of consciousness has been around for centuries, the fact there is no widely accepted definition is already telling us something: that perchance there is no definite definition or if there is, it will be always arbitrary, subjective, relative to some other factors. By the way, the same thing occurs with the notion of life that after so many centuries there is not one accepted definition of life, but this will one theme of Part II. Let us get back to consciousness. If finding a one-sentence definition is barely feasible and doomed to failure, then, can we accept to define it by enumerating its features? Could it be that the (scientific) study of consciousness becomes more reasonable if we do not insist on giving a precise and concise one-sentence definition of consciousness?

Defining phenomena through their properties is an accepted practice, just think of some fundamental concepts in science like linearity and nonlinearity, which probably some readers learnt in high school. These are defined by enumerating their characteristics: a linear system is defined by two main features, additivity and homogeneity (constituting the principle of superposition), and conversely, a nonlinear system is that which does not possess these two characteristics. It is also known that some fundamental concepts in science are nearly impossible to define. For instance, energy—while there are very specific definitions of energy that are associated with particular environments and conditions (so physicists talk about these types of energy: potential, kinetic, gravitational, electrical and quite a few more), there is no general definition accepted by physicists as of today. All attempts at finding a global definition of energy tend to be circular, hence useless. You can try the most common scientific definition of energy, the capacity for doing work, but work is defined in terms of forces, and forces are defined in terms of electromagnetic (or other) fields that exert a work, so we find ourselves at the start, and fields are energy after all, so there is no escape from the circularity. Consciousness (and life) could be another fundamental phenomenon for which no strict all-encompassing definition can be found. Let us then inspect the advantages of defining consciousness via its features, and we will see how some contentious questions about consciousness vanish one by one.

Some possible features through which consciousness can be defined—or better yet, described—include sensing (perceiving the environment, perception of the body), responding to stimuli, attention, memory, choice-making and perhaps the pinnacle of this (arbitrary) hierarchy could be self-awareness—self-evaluation or self-monitoring thoughts, the experience of agency. One can add a great number of features to the list, and hence the first poignant question that vanishes is whether it is possible to define consciousness: the answer is yes, by enumerating attributes just as was done with the definition of a linear system.

A second question many times discussed among philosophers and scientists that becomes evanescent is whether it is possible to study consciousness scientifically. This query could be surprising to some because, as was revealed above, there are currently many studies and conferences on consciousness, so someone must be studying consciousness. Yet, there have been influential scholars who have debated the question of whether or not consciousness can be empirically studied, this scepticism being due to the inherently subjective, personal nature of the phenomenon. With the help of our definition of consciousness the answer is yes, since each of those aforementioned features can be scientifically studied: as we have seen in this text, neuroscience is examining how we make choices, how we perceive the world, how we react to it, the levels of self-processing, memory, etc.

One more question that finds an answer with this manner of "defining" consciousness is that—many times asked—query about who or what is conscious; is a worm conscious? How about microorganisms like bacteria? A mouse? As we already revealed in Sect. 1.5 when discussing the cognitive powers of a worm, this so much-debated question can be partly settled because, say, a bacterium can sense and respond to the environment using its chemical senses and biochemistry in its tiny body (made up by just one cell!) hence it is somewhat conscious since it has two features of consciousness, perhaps the two most primordial features. Same with mice, these animals have more features, including memory retrieval and the ability to locate themselves in space. Therefore, without falling into panpsychism, one can admit that all living beings, including plants, display some consciousness attributes. Plants sense and respond to the environment too, have you seen the sunflowers tracking the sun? And you can check this paper if your curiosity is piqued about why sunflowers face the sun: Atamian et al., 2016, "Circadian regulation of sunflower heliotropism, floral orientation and pollinator visits", Science Vol. 353, Issue 6299, pp. 587-590 or in the direct link here[1]. Exciting research has also shown that trees communicate with each other and cooperate within the complex system of a forest (Simard et al., 2012). Thus, to some extent all these entities are conscious, but only the most advanced creatures have all possible attributes of consciousness. It is again a matter of gradation, remember that very few things in nature are all-or-none, rather a continuum is present at all levels of description.

[1] https://science.sciencemag.org/content/sci/353/6299/587.full.pdf

And finally, another question that loses importance with our "definition" of consciousness is about its location. It could be amusing to some to know that there was a time when the location of consciousness was thought to lie around the brainstem (about this brain region see Sect. 1.8.1), while other scholars proposed the thalamus and the cortex as the site of consciousness, yet others insisted it was somehow distributed in the large-scale integration of neuronal activity in cortico-thalamic and brainstem areas. Well, just as we noted when we talked about the location of the self that essentially the whole nervous system was implicated, the same occurs with consciousness. To review the scientific evidence that the widespread mutual neuronal activations among disparate brain areas result in conscious awareness is beyond the scope of this text, and interested readers can find a good number of studies on the topic. Suffice to say that if consciousness is described by those aforementioned features (and more that can be added), then it becomes apparent that it is a matter of the whole brain: perception takes place in several sensory areas devoted to touch, vision, hearing etc.; motor responses start in the motor areas of the frontal lobe; memory is stored in neocortical areas and let us not forget that the brainstem and other subcortical areas act as modulators, setting up "boundary conditions" for consciousness to take place. And looking at this from another viewpoint, that of removal of brain parts, if you remove the brainstem you lose consciousness, and the same happens if you remove your thalamus or your cortex (now, for those restless minds, let us say that a child was once born without most of the brain, only his brainstem remained, this being a rare condition called hydranencephaly where neural tissue is replaced with cerebrospinal fluid, and as a result of having only a brainstem—recall from Sect. 1.8.1 what this brain region does—he was able only to breathe and maintain a heartbeat, but little more, so he was mostly unaware of things, that is, unconscious; the child died at 12). The take-home message, in few and simple words, is that consciousness is not a static state but a process in which information is continuously transferred between various parts of the brain; the set of elements underlying consciousness is not static, but forms a dynamic core. Once more, we leave the stage to Roger Sperry who said "Consciousness [...] is a dynamic *emergent* of brain activity neither identical with nor reducible to neural events of which it is composed [...]. Not a mere passive correlate of brain events, but as active causal determinant essential to normal cerebral control" (R. W. Sperry, "Forebrain commissurotomy and conscious awareness", Journal of Medicine & Philosophy, 1977).

Then we seem to have answered in part that question about a location, but to more completely answer it, and to understand and naturalise consciousness and thus remove more of its mystery, we need one more concept. The keyword is that shown in italics in Sperry's quote above: emergent. It is worth mentioning a few words about emergence, because this is a most important concept in today science. From physics to biology or geology, emergence is considered a unifying paradigm. The next section covers the very basics of this phenomenon to see how it assists to understand consciousness.

7.1 On Emergence

A good and easy to read introduction to emergence is David Pines' article "Emergence: A Unifying Theme for 21st Century Science", published by the Santa Fe Institute. We all have seen the amazingly coordinated movements of a flock of birds (Fig. 7.1) or a school of fishes, or the symmetry of a snowflake, and soon you will see (Fig. 7.2) the hexagonal pattern in heated oil that will appear in your saucepan as you warm it up. These disparate phenomena all have in common that the collective behaviour of the whole is different from that of its parts—birds, fishes or oil molecules. *Emergence* thus refers to collective phenomena or behaviours in complex systems that are not present in their individual parts, because nothing in the molecules of oil indicates they should form a hexagonal array when heated, or a bird alone will not fly in that manner unless placed in the middle of the flock while making a turn. The pattern emerges as a result of the interactions in the group among its components. It is shown too in Fig. 7.1 the most frequent formation of a few birds during long flights, the V-formation (which is mimicked by the planes of the air force); this arrangement too is a property of the whole set of birds in flight, because it allows the animals to take advantage in the updraft created by the flyer in front so that the bird catches an extra lift, and this is why it is normally big-winged, large birds that fly like this, smaller birds cannot benefit from this energy-saving method because they produce tiny air currents while flying. Hence, the V-formation emerges from the interaction amongst the birds and their environment (how about the military aircrafts, do they fly in V-formation to use the same trick as the birds? Not quite, the main reason seems to be the angle of vision, although some calculations suggest the planes benefit too from air drafts).

A key point in emergent phenomena is that it is very hard if not impossible to determine a central control. Which is the bird that controls or starts the flocking

Fig. 7.1 Patterns in nature: emergence in action. A flock of birds flying in a coordinated fashion, and the typical V-formation of a few birds in flight. These are samples of emergent phenomena, the collective behaviour of the whole set of individuals that would be hard to anticipate if each one was studied in isolation, separated from the rest. The pattern emerges as a result of the interactions among the components of the group. The nature of the interactions will vary depending on what constituents form the system; in liquids there are molecules and thus there will be molecular interactions via, for instance, electromagnetic fields; in herds, the animals interact through their senses (hearing, vision, etc.) and in societies humans interact using a huge variety of means and these interactions result in sociodynamic patterns

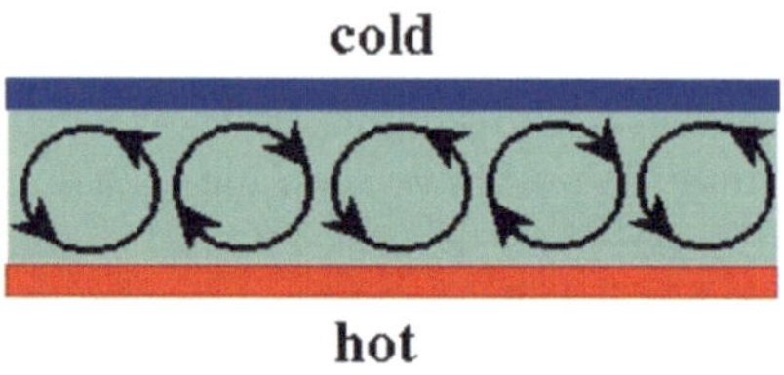

Fig. 7.2 The Rayleigh-Bénard convection in a heated fluid (silicone oil). An initially motionless fluid layer becomes unstable to flow perturbations when the temperature difference between the bottom hot and the top cooler fluid is sufficiently large, and convective rolls of millions of oil molecules materialise (depicted in the diagram), such that looking from above one sees a hexagonal array. This phenomenon is used here to explain the concept of emergence and a simile is made with consciousness as an emergent property of nervous tissue (see details in the text)

organised flights in Fig. 7.1? Or, what is the main control parameter that determines the hexagonal array in the heated oil in Fig. 7.2? We shall address this in the following paragraphs as this is essential because if we can conceive that consciousness is an emergent phenomenon of nervous tissue, then, what does this say about who is in control? We seem to go back to those ventriloquists of Sect. 2.2.1.

The subject of emergence is very wide and can become too technical for this text, so for the purposes of understanding how consciousness is an emergent property of the brain, we will use a paradigm of emergent behaviour, that by the way you can experience at home while making dinner: add a thin layer of oil (silicone oil works best but it is not recommended for cooking! You can see it with olive oil too) to a frying pan and heat it from below. In a few seconds, a geometric pattern on the oil's surface will be visible, shown in Fig. 7.2 (with olive oil the pattern is not that clear and you will have to watch carefully because as the temperature rises the pattern quickly disappears). This is the Rayleigh-Bénard instability—or convection—discovered by Henri Bénard (1874-1939), a paradigm to demonstrate pattern formation and emergent properties and also dynamical bifurcations (those into chaos theory have probably heard about bifurcations). For those interested in knowing why this happens, what follows in this paragraph is a technical note. The instability resulting in the appearance of the geometric pattern occurs because the initially motionless fluid layer becomes unstable to small flow perturbations when the temperature difference is sufficiently large—this was studied by Lord Rayleigh, and thus the name of the phenomenon pays homage to both scientists. Initially, when the temperature gradient between the top and bottom is not large, heat is transferred mainly by conduction—molecules bouncing against each other—and the fluid molecules remain at rest. As the temperature gradient between top and bottom grows, heat is better transferred by convection: the coherent motion of the molecules. Thus, small convective fluctuations develop, the hot fluid near the bottom thermally expands and becomes lighter than the fluid above it, rising, then cools down and returns in an overturning flow (depicted in Fig. 7.2). Some of the fluctuations become amplified and organised into a coherent motion, originating a flow pattern that consists of rolls or hexagonal cells, this pattern depending on the details of the fluid properties.

To summarise and extract the essence of the somewhat specialised previous words, the hexagonal pattern emerges from the combined actions of heat that creates a temperature gradient, the properties of the molecules in the fluid, and surrounding factors such as gravity (looking at the diagram in Fig. 7.2, it can be seen how the molecules go up and then go down due to gravity, setting up the flow forming those "rolls" depicted). As mentioned above, there is no central control in emergence or if there is, it is nearly impossible to ascertain what is in control. The phenomenon depends on those factors in an equal manner, such that if you don't heat up to establish a temperature gradient or heat it too powerfully and the gradient does not have the time to set up, then there is no pattern; in the same manner, if you change the fluid properties and use water instead of oil, you will not see any pattern; and if you do the experiment in deep space with almost no gravity, the convective rolls will not form and that will be the end of the assay. So, what is in control? No factor seems to be more important than any other... Other than the cook doing the experiment!

It can thus be stated that the dynamical repertoire of the fluid—that contains the possibility of creating the hexagonal configuration—is constrained by the range of microscopic molecular interactions and environmental perturbations, all of which determine the final dynamical evolution of the spatial arrangements. This sentence can be translated to neuroscience, to our main topic in this section about the origins of consciousness, and one can say that the dynamical repertoire of brain networks is constrained by the range of neuronal interactions and environmental perturbations, all of which determine the final dynamical evolution of the nervous system resulting in various behaviours. This sentence makes sense because in the case of brains we find too a multitude of internal and external factors—ion gradients, electrical and chemical interactions between brain cells, hormonal influences, sensory stimuli and a long et cetera of elements—that determine the final organised pattern of neural activity. Remember the—always overemphasised in this text—importance of the transient patterns of coordinated activity in a distributed set of neural networks as the cause of our behavioural dispositions, coordination of activity which normally takes the form of synchronised neuronal firing. And one final analogy between these two phenomena, if the coherent motion of oil molecules over macroscopic times results in that hexagonal pattern, the coherent activity among brain cells over macroscopic times brings about a pattern of activity that determines a specific behaviour; the specification of macroscopic times, meaning relatively long periods of time, is because these phenomena take some time to appear. It is not a matter of just a fraction of a second.

And just like in the Bénard convection the study of the oil molecules separately without considering other factors like gravity or temperature will not allow us to foresee the final arrangement, the study of the properties of single brain cells will hardly inform us as to the assortment of our possible cognitive states. And please do not take these words as implying it is useless to study in ever-finer details the intrinsic properties of cells. These are necessary studies, rather what was meant is that this sort of reductionist approach has its limitations when dealing with complex systems governed by emergent phenomena.

A query for those with excruciating curiosity: why hexagons and not circles or rectangles? A hint is now offered, bees also make hexagonal cells in their beehives to feed their progeny... But wait, these are two totally different phenomena, one belongs to the living and the other to the nonliving, and yet, related? Can the same principle solve these riddles? The answer in Sect. 8.1 in Part II.

7.1.1 A Criticism and a Reply

To be fair, we should mention a common criticism of this view of consciousness and self-awareness as emergent properties of an embodied nervous system immersed in an environment. Whereas considering consciousness from the perspective of an emergent phenomenon is an accepted idea by many scholars, one criticism that sometimes is voiced is that it does not help much to understand the mechanisms of consciousness. In other words, how all those specific matters we have discussed in this book so far—neuronal activity between this or that brain area, neural network interactions involved in self-perception, etc.—become more transparent by this idea of emergence. Is consciousness to the nervous system what digestion is to the gastrointestinal tract? After all, digestion can be thought of as the emergent property of the stomach (and both digestion and consciousness are needed for a happy life!), but this view does not seem to shed light on the specific mechanisms of the digestive processes. Yet, at the high level of description, this perspective has value in that it informs us about the global result of a myriad of particular processes and as well about the possibility or impossibility to access or investigate certain aspects of the phenomenon; and in any event, this view does not preclude us from investigating each of those minute and specific mechanisms associated with, say, digestion (how fat is absorbed, how sugar is metabolised and so on) or consciousness (sensory perception, etc.).

To fully reply to this criticism one has to reflect on the issue of levels of description to describe natural phenomena. The proposed view in this section is a high level of description, and then one can go to lower levels of description, say, neurophysiological properties of cell networks or of individual neurons. Figure 7.3 displays some events and elements that can be found at these distinct levels of description, from molecules and ions to whole ecosystems. Each level can be studied with different methods and techniques. To wit, brains can be examined across a spectrum of spatial and temporal scales that span various orders of magnitude, on one end of the scale there are single-neuron biophysical models and on the other end, at the whole-brain level, models of emergent network dynamics are popular among the members of the field called systems neuroscience. A high-level description that deals with a macroscopic level is not supposed to say much about the lower levels mechanisms, in the same manner as the study of these low-level, microprocesses may not reveal

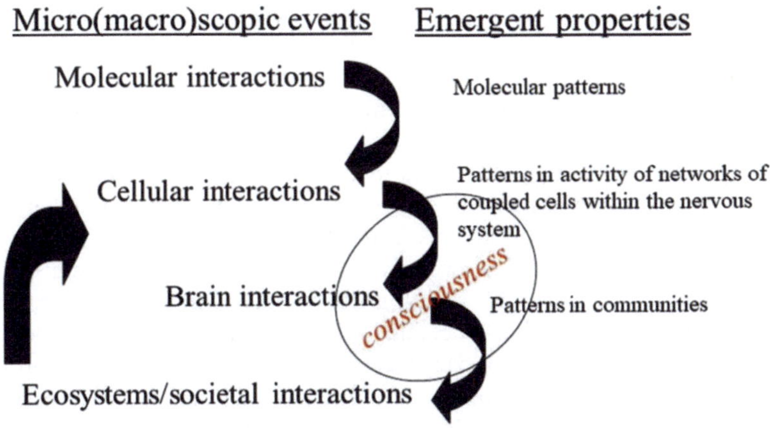

Fig. 7.3 Levels of description and the conceptual location of consciousness. Simple scheme illustrating some events at different levels of description and their associated properties in order to consider consciousness as an emergent property of the whole system derived from the local physiological characteristics, fundamentally interactions among cells in the brain, as much as the local molecular interactions in silicone oil give rise to the geometric patterns of the Bénard convection of the previous Fig. 7.2. On the left side, there are microscopic events at the start, that soon turn into macroscopic when many cells, organs (brains) or individuals in ecosystems are interacting. On the right-hand side some possible emergent properties, derived from the microscopic (or macroscopic) events at each level of description, are indicated. At the micro-level one finds molecular patterns, like that of Fig. 7.2, resulting from molecular interactions, and at macroscopic levels, there are interactions among individuals and groups of entities forming ecosystems (or human societies) which in turn affect previous levels (the thick black arrow pointing upwards), so, for instance, connections among people will affect the psyche of some individuals thus altering neural patterns in those brains. Consciousness can be conceivably located around the levels of neuronal (cellular) interactions and interactions among individuals, but please read the text to understand how all levels are interweaved

much about the higher-level behaviour of the system. One may investigate deeply the mechanism of how neurons fire action potentials in, say, nerves of the pain sensory terminals, but to more comprehensively understand how one feels pain one needs the high-level, or systems-level, description incorporating the findings from the pain sensory terminals to the several brain regions involved in processing those painful inputs. When you measure the temperature of your room using a thermometer, you are measuring an average of the velocity of the air molecules that determine the temperature (the faster the velocity, the higher the temperature), and the study of molecular motion can be done using theories developed by scientists like Maxwell and Boltzmann. So temperature can also be considered an emergent property of a mass of molecules, and nobody would say it is useless to measure it. The difference with the issue of consciousness is that we are not that advanced as in the field of thermodynamics, so we cannot measure consciousness as easily as we measure temperatures (well, this is not quite right, there are coma scales that sort of "measure" consciousness in patients in coma, but these very specific details are beyond our scope here), but perhaps in the future, there will be other techniques and theories that will link more deeply the different levels of description as it has occurred in thermodynamics or statistical mechanics.

Nonetheless, if we insist on abruptly mixing these levels of description with a brute force approach, it leads nowhere. As an illustration, it was already commented in Sect. 2.2.1 on the making of choices that there are proposals for the quantum world as the explanation of free will and consciousness. Well, it is true that, for all we know today, quantum physics lies at the fundamental level and all nature is based on quanta, but to try to describe psychological phenomena using quantum physics methods is not the best idea, these are two widely different levels of description. We should be content with inspecting each level using its methods, models and theories, understanding that from one level derives the next one and that laws governing one level become different laws at other levels, 'being and becoming' also applies here; and as well, we should have enough mental flexibility to jump from one level to another, without crudely amalgamating them but with the comprehension that, to continue with the temperature simile, what is the velocity of molecules at one level becomes a number reflecting temperature at another. In conclusion, the idea of consciousness as an emergent property provides us with the understanding that it is an attribute of the whole system—the brain which is coupled to the rest of the nervous system (and please do not forget the body, remember the embodied nervous system)—and does not reside in a specific localised brain area, at least for all we know at the time of this writing.

And along this line of reflections about levels of description and the location of consciousness, let us finish with the suggestion mentioned at the end of the past section that in order to finalise the idea of "locating" consciousness, the considerations about complex systems and their associated emergent features would be of help. Therefore, as a manner of a summary, Fig. 7.3 portrays a possible conceptual location of consciousness and self-awareness.

An important point to consider is exemplified in Fig. 7.3 by the black arrow pointing upwards from ecosystems to previous levels from where these originate: in ecosystems or societies there are interactions among individuals which in turn are determined by cellular activity in each individual and so on back to the subatomic, quantum level if one wishes. This is a depiction of the fact that emergent patterns at each level act on and modulates/constrain previous levels. In the parlance of physics, the order parameter feeds back on the control parameters. By the way, the entities that interact need not be as material as atoms, cells or people, these could be the symbols with which at a certain, cognitive level, the brain operates, the symbols being representations of some aspects of the reality experienced by the organism that has a brain; so for example intentions or goals, being high-level cognitive events or "symbols", can be conceived to act as constraints of the brain dynamics that will result in certain actions. For those with a tendency towards a holistic state of mind, the world of emergence has a certain beauty.

Then, after all these considerations, what can we say about Colin McGinn's words that "Consciousness [...] must be a natural phenomenon naturally arising from certain organisations of matter" which appeared in his article "*Can we solve the mind-body problem?"* (McGinn, 1989). Are we ready to explain how consciousness arises from the organisation of matter? What can be concluded from a high-level perspective is that the neuroanatomy which determines connections with sufficient complexity among brain cell networks sets constraints to the connectivity patterns, and at the same time the neurodynamic laws—from which brain cells derive their activities—define functional, or meaningful, interactions among cell networks. From this set of complex and organised interactions, cognition and all its aspects emerge. Once more, connectivity patterns specifying organised, coordinated cellular activity are the key to comprehend how from a soft, mushy tissue like that of nervous systems, cognition, consciousness and related phenomena materialise. The investigation of those neural organised activity patterns is under close scrutiny in contemporary neuroscience.

7.2 Perception is Personal: The Essence and Source of the Mystery

To end this chapter and part I, it is honest to say some words about the main reason why consciousness is seen as something mysterious, which somehow was not treated in the previous paragraphs. This reason lies in that consciousness is a subjective experience. Right at the beginning, in Chap. 1, we saw that perception is everything, and perception is a very personal affair. The importance of perception for the existence of the self was established in Chap. 5. In the final analysis, all consciousness is consciousness (or awareness) of something, or, correspondingly in Antonio Damasio's words, consciousness is the feeling of what happens—just try to imagine yourself unconscious, and this is not to imagine yourself lying quietly in bed, it is yourself without any perception, can you experience unconsciousness? This is an apparently self-evident allegation but that is of essential significance because it puts all emphasis on the nature of perception and how we (our brains) interpret those percepts, recall Fig. 1.4 in Sect. 1.1. Hence, we are dealing with obviously a very personal, subjective business. And because the feeling, sensation or experience, of those percepts feels (yes, a bit redundant, sorry…) like immaterial to us, then the question that has been asked countless times through centuries is how the immaterial experience of consciousness can be explained by the material cells or tissue of the brain.

Nevertheless, some scholars argue that "although subjective experience is unique in nature, it is not necessarily mysterious", these are the words in the overview of the book by Feinberg and Mallatt mentioned at the start of Chap. 7—*Consciousness Demystified*—where they endeavour to do in a more technical and specialised fashion what we have tried here in a few pages and in simple words: naturalise consciousness, by explaining that the subjective, experiencing aspects of consciousness are created by natural brain processes that evolved in natural ways. The look at the uninterrupted progression of cognition, the realisation that there is a continuum in the evolutionary development of consciousness from worms to humans, the comprehension that adding complex and intertwined neural circuities to primitive nervous systems allows to go beyond the pure reflex and results in more cognitive power, all these things that we have tried to show in previous sections help remove that feeling of mystery. Still, some scent of enigma remains, there is something in our subjective experience of perceptions that leaves us wondering whether it is legitimate to ask about the bridge between these subjective feelings and the physical substrate of the nervous system.

As it turns out, these instances of subjective conscious experience have been given a name: qualia. This is a favourite topic of discussion among philosophers of the mind. In plain words, they debate things like the nature of the sensation of the colour red feeling like red. This issue of explaining the subjective nature of consciousness arising from its material substrate is popularly known as the "hard problem", which has kept, and still is keeping, philosophers and some scientists busy. As this is not a philosophy text we shall skip these discussions, and will only mention that some scholars think that the hard problem is empirically intractable (Cohen & Dennett, 2011), and also what the philosopher Merleau-Ponty, whom we met before when discussing free will, said about the quale: "sensations are far from being reducible to a certain indescribable state or quale, they [...] are enveloped in a living significance" (Merleau-Ponty, 1945). Those readers willing to face his voluminous book on the phenomenology of perception will appreciate his efforts at naturalising experience, and for those who do not wish to take the trouble, we add here a few straightforward words summarising his view, that our mind noticing the body's reaction to the world and responding to that experience constitutes the roots of consciousness. His philosophical ideas, with the insistence that without our bodies there can be no consciousness, were part of the stream of thoughts that developed into what is known as neurophenomenology, whose principal proposal we have already seen here a number of times, yes, it is that embodied condition of the human mind, the embodied brain immersed in an environment. Further readings on the topic can be found in an article by one of its founding progenitors, the neuroscientist Francisco J. Varela, "Neurophenomenology: a methodological remedy for the hard problem" (Varela, 1996) and in Gallagher, 2009. And with this, we close this paragraph on the "indescribable state" called quale.

It has probably transpired that the perspective favoured in this text is similar to that of Merleau-Ponty in that it may not be necessary to invoke things impossible to describe, in spite of the subjective nature of perception and experience. The view of other scientists like Edmund Rolls may be more pertinent to naturalise experience: "Consciousness may be the state which arises in a system that can think about (or reflect on) its own thoughts, that is in a system capable of second-order thoughts [...] If a system were doing this type of processing (thinking about its own thoughts), it would then be very plausible that *it should feel like something to be doing this*" (Rolls, 1999). And of course, here we go again "... should feel like something ...". We can thus see that trying to describe or define this matter leads us to circular descriptions, using notions like feelings or sensations to describe perceptions and the associated sensations or feelings; this circularity already is telling us something:

Fig. 7.4 The (expanded) window metaphor of consciousness. Gloor's window metaphor explaining consciousness is described in the text. Here we take the liberty to add one more part to it, the window frame as "containing" the aspects of consciousness used to perceive and interpret the view. Please see the text for an explanation

as was mentioned in the first paragraphs of this Chap. 7, attempts at finding a global, general definition of fundamental properties of nature like energy or matter tend to be circular, hence useless. Interestingly, some philosophers of the mind argue against that persistence in describing subjectivity, as for example, John Searle who admitted that "one cannot represent subjectivity as part of our representation of the world because subjectivity is nothing else than this representation itself" (Searle, 1992).

In a similar vein, the thoughts of Swiss neurologist and neurosurgeon Pierre Gloor provide us with a simile that can be expanded to explain disorders of consciousness, depicted in the cartoon of Fig. 7.4. Gloor's window metaphor as an explanation of consciousness appeared in his 1986 paper and note that Searle would agree with him: "Consciousness cannot, therefore, be external to itself; it cannot be an 'object, out there'; it thus cannot be observed. If I may be allowed to use the metaphor of describing consciousness as the only window through which we can look at the world, then it follows that when looking through this window, we cannot see the window itself, even though it, too, is part of the world. Consciousness thus conceived is not an objectively verifiable datum; it, therefore, cannot be defined, and its very nature is not accessible to any form of objective analysis" (Gloor, 1986). So you see, Gloor agrees with the vision presented in this book, that it is best to forget about finding a strict definition of consciousness, but with regards to not being "accessible to any form of objective analysis", that may not be quite accurate because, as already explained above, one can study using "objective analysis" the many features of consciousness.

In Fig. 7.4, we have the upper window with a frame that represents or incorporates the features of consciousness needed to perceive and interpret the view through the window. This frame was not originally in Gloor's description but it could be an illustrative addition to the simile. Accordingly, then, consciousness is that cavity, the window through which we look at the world, but since it is empty, we do not perceive it but we perceive the world outside; we cannot see or study the window itself, but we can see and study the frame, those characteristics of consciousness that were mentioned in our proposal for a definition of consciousness by enumerating its features. In the lower cartoon, we find the same window with a broken frame, representing faulty features (bad memory, defective sensory organs, and so on) that may be present in different patients, which will prevent a true and complete view of the world through the window. This is then the expanded Gloor's window metaphor. Just one more thing about Gloor's thoughts, despite our disinclination to define consciousness in one sentence, if we were to choose one, we would probably agree with his definition that appeared in that same 1986 paper, "Consciousness is a unified experience with a measure of continuity in time and with an invariant central point, the self".

As final words on the subject of consciousness and before we take on the matter of life in Part II, it is to be noted that underlying the disputes and the different views on consciousness is the fascination of people with making dichotomies, with separating phenomena that can hardly be separated. The mind-body dualism is an ancient one indeed, separating mental from physical states. Lurking here is the tendency to establish—and subsequently believe in—clear demarcations between phenomena, without considering the most fundamental point that nature normally works in a graded fashion, a sort of continuum. So questions like 'when does consciousness emerge from mere perception'—or its equivalent that we shall address in Part II, 'when does life emerge from mere chemistry'—lose part of their significance after understanding that perception is already a feature of consciousness, therefore there was not an abrupt step that occurred during evolution and suddenly an animal became conscious because, as we have explained in this Part I, all living entities have features of consciousness and thus are to some extent conscious. Another question derived from another dichotomy: Does the brain create the mind? One way to grasp the fallacy behind this dichotomy is to think that this is like asking whether the body (material substrate, akin to the brain) creates life (man-created concept, akin to mind), although for some reason nobody asks this question but many ask the former. Life and mind are our arbitrary, abstract notions, brains and bodies are the real physical stuff. These notions help us—scientists, philosophers and people in general—to talk, to communicate in a relatively straightforward fashion, but problems arise if we start to believe blindly in these concepts and ideas. We have to take them for what they are, products of our intellects, arbitrary concepts in the final analysis. In the words of Nobel Prize-winning biochemist Christian de Duve "We must accept our concepts for what they are, provisional approximations that are as much fictions of our minds as they are faithful depictions of the facts". Finally, a dichotomy is also lurking in the aforementioned qualia business. We normally say "I had a feeling of pain", but not "I pained". Why? What's the difference between the feeling of pain and pain itself? None, if you think about it. We feel pain because our nerves/brains can interpret the event as what we call pain. Does a tree falling in the forest make a noise if nobody is around to hear the fall? Same thing here, noise/sound is airwaves, but you need an organism that interprets those waves as sound. So sound and pain exist independently of the interpreter, and at the same, they need the interpreter to be interpreted as such! So that quale we talked about is a way of saying "privatisation of the sensory experience", when perception combines with other contents of the brain (memories…) and, in a sort of reality check that brains continuously perform, the subjective experience arises and as its name indicates, it is not an objectively verifiable datum (using Gloor's words above). And naturally, this privatisation of the sensory world occurred when the evolutionary progression allowed brains to possess the mental activity needed for such a thing to occur.

Furthermore, if consciousness arose as a consequence of reaching certain organised complexity in embodied nervous systems immersed in an environment, then the location of consciousness shown in Fig. 7.3 reflects the global nature of it, needing not only brains and bodies but also interactions among those bodies in the ecosystems (or societies). It is in this sense that consciousness is a global property, not of one only brain but rather of the whole ecosystem. And there is no need for mysticism or deep mysteries here, it is a rather natural phenomenon based on the physical substrates of the universe, principally the interactions among the constituents. There is nothing esoteric about this.

7.2.1 *Postscript to Part I*

To recapitulate some key points of Part I, we have seen how the emergence of the self—and equally the rest of the features of consciousness—is an inevitable consequence of our brain, body and situation in the world, and that this sense of personal identity has advantages for the organisms that have it. The self helps substantiate the psychological need for agency, people need to see themselves as agents causing things around, experiencing reality, and this need results in the seeking for that sense of personal identity that, in some individuals, become almost an obsession, nonetheless, in more or less degree it is present in all of us.

But we have also seen how this sense of self is a model, a construction or illusion of our brains that is updated according to circumstances. The neuroscientific evidence teaches us how the sense of personal identity is a matter of perception; depending on how the body and events are perceived or depending on what neural networks are more active, our sense of agency changes, the self-model is altered. We have gone over several neuroscientific observations in patients and animals that should start to cast some doubt about our sense of self and our perception of reality in general, as indicated in Unamuno's words, already presented at the start of our text, "The supreme triumph of reason is to cast doubt on its own validity". So part of our triumph is the understanding of the reasons we develop a sense of self, of what that personal identity is and how it becomes—how it changes, how the model is updated—, understanding and accepting it and equally admitting that the rest of things around us are and become something else, how events progress. It is fine to try to understand oneself, but the reflections in the above sections should warn us that becoming very obsessive about this self-understanding and being excessively preoccupied when one cannot make sense of him/herself, is not the best strategy to lead a happy life, if only because, as became apparent in this part I, the more one tries to find that elusive self, the more it seems to retreat into the crannies of our minds. If there is a global take-home message, this would be the acceptance of our personalities as these are, trying perhaps to improve aspects that make us and others unhappy, and similarly, the acceptance of the alterations that will, most likely, occur in the future.

One aspect worth mentioning after so much talk about ventriloquists in the brain is that there are others throughout the body. Since the nervous system is connected to everything else in the body, other organs will influence those ventriloquists in neural tissue competing for power. Take for instance the mutual interconnections between the immune and the nervous system, but more interesting is perhaps the consideration of the so-called second brain: the gastrointestinal tract. The gut is inhabited by billions of living creatures, microorganisms that to some extent can dictate our mood—think about how you feel during heavy indigestion. We are not alone; we carry a myriad of passengers in our trip through life. If we look within ourselves, each of us will recognise that we contain and are enveloped by a universe of microorganisms. Our largest organ, our skin, is covered in microorganisms from four different phyla that coexist (Grice & Segre, 2011), and can differentiate between the beneficial microorganisms and the pathological that can cause disease. Our mucous membranes are host to other species, and our gastrointestinal tract denizens have collectively been referred to as a second brain. The gut microbiome in particular pays its rent in producing neurotransmitters such as serotonin and then sending signals to the brain via the vagus nerve. This two-way highway is a reciprocal information exchange with the gut bacteria influencing the brain and in turn being influenced by the same hormones and neurotransmitters produced by the host. Alterations in the microbiome are associated with a variety of chronic "physical" and "psychiatric" conditions —the quotation marks again highlight the artificial boundary between body and brain. For instance, irritable bowel syndrome is influenced by dysbiotic states (Bull & Plummer, 2014), as are depression and Parkinson's disease (dysbiosis means a reduction in microbial diversity and the loss of beneficial bacteria). While we are not conscious of the multiplicity of other "selves" cohabiting in and on us, they are integral to our existence.

Part II will be devoted to the other search mentioned in the initial words of the book: the search for a purpose in life, for a meaning of life. If understanding what our selves are and how they develop help us alleviate to some extent some burdens, understanding of the incessant becoming in life will aid to clarify why we came into being and for what purpose, if any. So bit by bit, we are finding some liberation from the fetters of existence in our bewildering world we have created. To begin comprehending life's becoming, we will start by considering an apparent miracle that will shed light on how complex and apparently organised patterns emerge out of no pattern.

Part II
The Biophysics Perspective – What Is My Life?

Certainly, Queen's musicians are not the only ones asking "does anybody know what we are living for" —in their famous song "The show must go on", perhaps an adequate music background for this Part II.

"The need to see 'purpose' in evolution, or at least some internal drive to help the blind processes of random variation and natural selection, is remarkably resilient", these are the initial words in the short paper by W. J. Dickinson and J. Seger on cause and effect in biological evolution. Indeed, many people strive to find purpose in life, and whereas for many this search is just a sort of distraction or an amusing query, for others it becomes a heavy burden, and the psychological side-effects of this search should be reckoned with. These authors also mention a bit later in the text that "But selection lacks foresight" (Dickinson & Seger, 1999). In the opening words of our book it was declared that the search for a purpose in life occurs in parallel with seeking a sense of personal identity, and that this text intends to reveal the fallacy in these searches. Part II is devoted to the understanding of the "purpose" in life. To understand the purpose of something, it is wise to study it, to know how it originated, what its main features are, and this is the reason for the sections that follow.

If in Part I it was shown that the appearance of the self was an inevitable occurrence, in this part we shall see that life too is an inevitable phenomenon in a planet with the characteristics of the Earth. And as it happened in Part I on the neuroscience of the self, some science – namely physics – will be needed here, but again the level of description will hopefully be easy enough to be understood by most non-specialist readers, because actually "Most of the fundamental ideas of science are essentially simple, and may, as a rule, be expressed in a language comprehensible to everyone", said Albert Einstein, and if you bother looking at many essential concepts in science and at the most fundamental equations of physics, you will see that, with few exceptions, these are really simple.

Before delving into details, let us start by exposing the main fallacy about the common view, held by many, which claims that we humans have evolved in such a precise and wonderful manner that it would have been tremendously improbable to have arrived at our quasi-perfect physiological state (but for some diseases of course) if it had occurred by chance, hence there must be something or someone behind our evolution dictating our past, present and future. Along these lines, some readers will possibly have heard or read about how unlikely it is for a protein in our body made up of, say, 100 amino acids to have that particular sequence of amino acid; therefore, the protein must have had a designer to carry out whatever physiological function in our tissues – proteins are formed by specific sequences of the molecules that constitute proteins, the amino acids, and the sequence determines the function of each protein in the body, and if you bother calculating the probability to find one specific sequence of 100 amino acids, you will see it is extremely low. Well, this notion of our presumed directed evolution rests on the concept of probability, which is relative, and it depends on the perspective. If you have one lottery number and there have been one million numbers sold, your probability to win is one in a million, that is, quite low; yet, from the viewpoint of having a winner, it is guaranteed one number will win, and hence, it is certain – maximal probability indeed – that there will be a winner. Then, what is it, low or high probability? Depends on what perspective you take. In the lottery of biological evolution, there were many numbers, but one had to "win". We, and the rest of animal and vegetal forms, are the winners of this lottery whose currency is survival. That is all.

Some readers may have been intrigued about the previous assertion that probability is a relative concept, so considering how many times we invoke this notion in our everyday situations it would be fruitful to recognise that what may seem sometimes very improbable on a cursory reflection, it is in fact very probable. Thus, let us start explaining why things, events, happen, the fundamentals of pattern formation in nature around us, where we will see how sometimes the improbable is in reality very probable if not unavoidable. The best way to start grasping this is to begin by performing a miracle: let us accomplish what seems an extremely improbable event.

Chapter 8
A Recipe for Interesting Things to Occur

Contents

A recipe for high cognition was offered in Part I (Chap. 6), and now, we present another for events to occur in nature, especially "interesting" events, some of which seem to be counterintuitive like the emergence of organisation and order out of disorder. Understanding this will improve our understanding of how life appeared, will dispose of part of the mystery of life and will broaden the comprehension of what it is all about, knowledge that may help some not to wake up every day with the intention to confront life, rather to flow along the course of events that constitute that life.

Now, to perform that aforementioned miracle, a box or container is needed with two separated compartments and a hole communicating them, plus a bunch of little beads that can go through the hole. Understanding the reasons for this apparent miracle where order will appear out of disorder facilitates to start comprehending the nature of phenomena. The experiment dates back to 1996 when it was described (although in a bit different fashion) by H. J. Schlichting and V. Nordmeier, but what follows is the version in Troy Shinbrot and Fernando J. Muzzio's 2001 article. The experiment starts by distributing the beads (steel balls work well) in equal numbers in both compartments of the box, and then this container starts to be shaken. In the authors' original words: "A number of steel beads were initially distributed uniformly across a vibrated acrylic container. The container is separated into two identical chambers by a foamboard barrier containing an open window near its bottom. Despite being distributed uniformly at the start, after a short time (2–3 minutes) most of the beads migrate spontaneously to one chamber or the other. This separates the system into a first chamber containing a sparse and gas-like state of high-speed particles, and a second containing a dense and close-packed state of nearly

J. L. Perez Velazquez, V. Nenadovic, *Being and becoming*,
https://doi.org/10.1007/978-3-030-78264-1_8

motionless beads. This separation of 'hot' and 'cold' states occurs in the presence of noisy agitation in the form of a macroscopic analogue of Brownian motion" (Shinbrot & Muzzio, 2001). Technical note: Brownian motion is how physicists called the random, erratic motion of microscopic particles, and it is named after the botanist Robert Brown, who, in 1827, described the motion of pollen particles while looking through a microscope at the pollen of a certain plant immersed in water; later on in the early 1900s—that is, about 80 years after Brown's description, so you see things work sometimes very slow in science and interesting observations may go unnoticed for a long, long time—the famous physicist Albert Einstein studied this motion and modelled it using mathematical equations, which became a convincing evidence that molecules exist.

In even simpler words, the particles/beads that are initially distributed in equal numbers in both chambers are free to travel between the chambers at all times through the little hole due to the shaker that vibrates the box thus providing energy to the beads making them move, colliding randomly; and it is due to the inelastic collisions between the beads that they migrate from one to another chamber through the hole and tend to remain in whichever side randomly acquires a slight excess of beads, because each time a bead suffers a collision, it loses (kinetic) energy owing to the inelasticity of the bead material (that's why steel works well), and once that a small surplus, by chance, of beads appears on either side of the barrier, that side will lose energy more rapidly than the opposing side because there will be more beads colliding and losing energy and therefore the beads in this compartment will move more and more slowly and will not be able to jump through the hole. At the same time, fewer beads on the other side will make them more energetic as they will experience fewer collisions due to the smaller particle number, and therefore, this chamber will continue to lose increasingly more particles that jump via the hole to the other side which will progressively contain more beads making the collisions among them increase and hence they will continue to lose more and more energy and become almost immobile in the end. The final result is that the disordered and homogenous initial state when equal number of beads were distributed in both chambers has been transformed into an ordered and heterogeneous state with almost all beads in one compartment, and all as a result of agitation due to the energy provided by the shaker and the loss of energy due to the random collisions among the particles.

Hence, we have witnessed an apparent miracle: something that looks very improbable has taken place, the beads spontaneously filling one chamber after an initial homogeneous distribution. This sounds like the air in your room suddenly concentrating on one corner and leaving the rest of the room empty of air molecules—but do not worry, this will not occur because, among other things, nobody is supplying energy to the air molecules as the shaker supplied energy to the beads in the box (well, to be totally precise, this event may occur but the probabilities are in this case truly infinitesimal, and you would have to live many lives to see that abovementioned spontaneous change in the room's air; few things in nature are truly impossible, but some are so improbable that in order to facilitate our straightforward communication they can be called impossible).

Just like in Part I when we talked about the unreliability of perception, we see here that events happening around us in the physical world may not proceed as we expect, and things that look very unlikely are in fact very likely to occur provided some characteristics are present. This is a hint, a very brief introduction, about what will be talked in Chap. 9 on how life appeared. The apparent paradox in this experiment—that the spontaneous formation of patterns looks implausible but, in the end, it is highly probable—is solved once we take into consideration two main features: the abovementioned energy exchange between two beads colliding and the source of energy supplied by the apparatus that shakes the box. We, therefore, are already peeking at some main features for things to develop in the world around us, for patterns to emerge in natural phenomena, namely energy supply and energy exchange through interactions. But we still need at least one more feature to complete the recipe for interesting things to occur.

In technical terms, this phenomenon of energy lost after two beads collide is called "energy dissipation", so in these scientific terms, a small excess of beads in one of the two chambers produces a large excess in dissipation in that chamber and will therefore be able to eject fewer beads than it receives. A system that dissipates—uses—energy is called in the scientific parlance a dissipative system. Our bodies are good dissipators because we transform food into energy and use part of that to maintain our homeostasis, and our jobs—perform our work. Energy dissipation is one factor determining the variety of natural phenomena in the universe: energy is transferred from one place to another, and this is, basically and in simple words, why things occur; the transition from being to becoming, the very basics of pattern formation in nature from lighting in thunderstorms to our brain's activity that, as we saw in Part I, determines our behaviour, is due to the gradients of energy. Therefore, we are now in a position to provide the recipe for interesting things, patterns, to occur, that is, for events to develop in our universe. If we take one of these ingredients of the recipe out, only boring events, not much in fact, will take place:

BASIC RECIPE for a lively universe: energy dissipation, nonequilibrium/nonlinearity and a source of noise.

This very basic recipe will help us understand why life, as we know it, has not much of mystery in it. But one step at a time. Let us explain a bit more, using simple non-specialised words, those three ingredients of the recipe.

8.1 Preparing to Cook Life

The recipe contains three main ingredients, energy dissipation was already seen in the experiment described in the preceding paragraphs about the creation of the arrangement of steel beads in one side of the container while leaving the other almost empty. Dissipation of energy is just a scientific manner to say what everybody knows, that energy is constantly in motion moving from high energy places/states to lower energy ones, energy is exchanged moving down these gradients. Of course, this requires an input of energy from somewhere, or in technical words that the system be open, able to exchange energy (and sometimes matter) with the surroundings. In that experiment, it was the shaker that provided energy to move the balls in the chambers of the container. Without energy, nothing can be achieved, or more properly, without energy gradients. If there is no gradient, that is if energy is equally distributed everywhere, not much can occur—think in terms of a metal bar where there is the same temperature along the rod, without a temperature gradient heat "cannot move" from one part of the bar to another. And naturally, these processes of energy moving down gradients, from high to low, result in the disappearance of those gradients as energy will tend to become equally distributed everywhere as much as the temperature in that rod above will become the same along the length after a while. What this means for the future of the universe is that if there comes a time when energy is uniformly distributed throughout the whole space, then what some have termed the heat death of the universe—also known as big freeze or big chill—will come to pass: no events will occur, there will be a "being" but not more "becoming". But do not worry, according to theory, the time it will take for the universe to stand still is around a googol of years, that is, 10^{100} years, (this excessively large number deserved a name that was coined in 1938 by a nine year old boy nephew of a mathematician, and has no practical use as it was originally proposed to illustrate the difference between an unimaginably large number and the infinite, and if you think the name Google has anything to do with the googol, you are right). Hopefully, this discussion has served for something more than knowing where the name of a search engine comes from, it has shown that if energy is provided to the constituents of a system and these are allowed to interact and exchange it, then things will occur. But these "things" will need to have a certain character to be part of what we term life, and for this we need the other properties in that recipe.

8.1.1 Global Regularity From Local Mess: The Beginnings of the Living

The next ingredient of the recipe is nonequilibrium (we can forget about nonlinearity for the time being lest we get into very technical details). In truth, this can be considered a consequence of the previous ingredient because if energy is provided to a system in equilibrium, normally this equilibrium will be displaced, and the system may even be sent to the known as far-from-equilibrium situation, depending on how much energy it receives. The world of far-from-equilibrium is in fact of tremendous importance in science, and the field of non-equilibrium thermodynamics teaches how patterns emerge in nature and how order may arise from disorder. So now we come to the concept of equilibrium, that while many of us learnt it in high school—remember your chemistry class?— it is a source of misunderstanding even among us, scientists. For this reason, let us clarify some points because this notion is a crucial one in understanding life, how we came to be, and as well the rest of natural phenomena.

In lay terms, equilibrium is understood as a situation when nothing changes. Same in science: when the properties of the system remain constant (there may be small fluctuations around the average), the system is at equilibrium. The fridge thermostat maintaining a temperature around 5 degrees makes the inside of the refrigerator a system in equilibrium. And if you measure your body temperature or your heart rate every 5 minutes while comfortably sitting down, you will see your body is at, or near, equilibrium. But now take a look at tiny pieces of your tissues, explore this microworld and you will see a very different picture, nothing resembling constancy or equilibrium. For instance, were you to look at your neurons under a potent microscope, you would see molecules and ions going in and out of the cells all the time; harking back to Sect. 1.4, it was described that neurons fire spikes or action potentials, which are electric currents therefore charged atoms and molecules—ions—need to move between the intracellular and extracellular compartments to generate those currents that will allow cells to communicate. Or, taking a microscopic look at hepatic cells of the liver you will see an intricate mesh of biochemical reactions where all molecules seem to be colliding with all, with some appearing while others being consumed. This does not look like equilibrium at all. How come from this molecular, microscopic pandemonium, that (macroscopic) order and equilibrium take place?

The answer to this question starts by considering what was presented in Sect. 7.1.1 about the different levels of description, where it was emphasised the fact that from one level, say the microscopic or molecular level, the next one is originated, and the properties and laws governing one level become different at other levels. Thinking on these terms and considering also the description of emergent properties in Chap. 7, it is not difficult to figure out how the mess and disorder of the microworld yield organisation and order in the macroworld. We witnessed it in the case of the emergence of the hexagonal pattern on the surface of the liquid in the Rayleigh-Bénard convection; this long-range macroscopic order is driven by local microscopic interactions among billions of oil molecules. And yes, talking about dichotomies again, this could be another one: the local versus global levels of description. But in reality both levels are two sides of the same coin. Although locally, in small distances or microdomains, many of our biochemicals are constantly changing, globally our bodies are at near equilibrium, which is called homeostasis: temperature, blood pressure, glucose levels and so on are kept constant, hence at equilibrium. A dead body will be too at equilibrium, perhaps a more complete equilibrium because in this case even the microworld will be equilibrated, will be kept in order as no energy will be used anymore to unbalance the uniformity, all gradients either energetic or molecular will vanish in the corpse, and that will be the end of pattern formation in the body: no patterns—biochemical reactions, electrical fields—, no life. Have we concluded then that a dead body and a living one are at equilibrium? Looks like it, but equilibria of different nature.

On this topic of equilibrium, it is fair to note that this concept is somewhat relative. One sample of this relativity we have just seen in the preceding paragraph is about the local and global views—those with a craving for knowledge can consult the somewhat technical paper 'Taking thermodynamics too seriously' (Callender,

2001) to realise how tremendously arbitrary this concept may be and as well to learn about other typical thermodynamical misinterpretations.

Equilibrium is an arbitrary notion that depends on the context and the level of description. Disregarding context and descriptive level leads to multiple misunderstandings in the application of this notion to natural phenomena. For starters, equilibrium depends on time intervals, it is related with the observation time. Imagine you take your temperature or heart rate not every 5 minutes as you did two paragraphs above but twice a day every 12 hours, at 3 in the morning and at 15 hours. You will notice in this case a larger variation than in the previous sampling in small time intervals (because during sleep the body temperature drops), which may lead you to think your body is not at equilibrium. The same impression you would have if you record your brain activity during wakefulness and later on during deep sleep at night because during the day the recordings are small amplitude wiggly lines, whereas at night, particularly during slow-wave sleep, your brain will show those slow, rhythmic high amplitude waves, little in common with the previous recordings of daytime; but taking the recordings at small intervals of, say, 10 minutes during daytime hours, all the neurophysiological traces recorded will look alike; hence, you will think about equilibrium now. How about a nonliving example: add boiling water to a cup and measure its temperature several times only during the first 10 seconds, you will have equilibrium because basically it won't change; but now measure it continuously for 10 minutes, it changes a lot, hence you have gone from the happy realm of equilibrium to that of non-equilibrium. But wait, continue measuring the temperature for the next 10 hours, and then you are back at equilibrium because it won't change once it is cooled at room temperature. So, is our world at equilibrium or not? It seems that depending on what frequency one measures properties, sometimes it will look that the system is at equilibrium and sometimes it will not.

Not only the observation time scale but also equilibrium is connected to the level at which one observes the system, as in the aforementioned difference between microscopic mayhem and macroscopic order. Even the spatial extension where one takes measurements will change our views: take the air pressure of the atmosphere at some points around your house and you will find the same value, hence equilibrium, now take it at 1000 kilometres away and you will most likely find a totally distinct value. If, accordingly, phenomenological equilibrium is intimately tied to an observation time-scale, to a spatial scale and to the level of observation/description, then, is it of any use? It is, in fact, much of classical thermodynamics is based on this framework around the equilibrium.

Why devote so many words about equilibrium here in this book? It is a somewhat specialised notion used in fields like thermodynamics, nonetheless, it is a concept that is many times misunderstood and misleading—unless the previously mentioned considerations are taking into account—resulting in wrong or inaccurate conclusions about living phenomena, our main theme here. Things are a bit easier with the nonliving and, moreover, if the system is a closed one, that is, an isolated system that exchanges neither matter nor energy with the exterior. So, if you are fond of chemistry, you can conduct a chemical reaction in a test tube, place it in ice, and wait until equilibrium—in this case true equilibrium—is reached; the molecules involved in the chemical reaction in your tube will remain constant over time. Not much will occur after equilibrium is reached in the limited universe of this test tube. But living processes are open systems which, at the same time that this complicates matters for their study and analysis, it is the source of the variability needed for complex, living processes to take place. Because these mechanisms are constantly exchanging energy and matter true equilibrium is almost never reached, rather they maintain a sort of steady state of non-equilibrium, just like our bodies in the examples above. Hence, in nature, we almost never see pure equilibrium, rather the approach towards it. The essence of the notions of the approach to equilibrium—and the easiest manner to comprehend what this means—is to think about it as the approach to the most probable state.

Furthermore, this richness of non-equilibrium situations manifested in the origin of different behaviours or patterns in nature has one hidden advantage for the student of life, namely that studying these processes from a high-level perspective—the one we are maintaining in these sections—we can forget the specific, particular mechanisms and focus on the final pattern or behaviour of the system because different mechanisms may give rise to the same phenomenon. As stated by D. Sornette, "the richness of out-of-equilibrium systems lies in the multiplicity of mechanisms generating similar behaviours" (Sornette, 2000). We are here seeing the very fundamentals of natural phenomena that give rise to almost all-natural patterns: allow interactions that exchange energy among the components of the system and provide some non-equilibrium situation including some noise or fluctuations, and you will get an enormous variety of natural patterns in front of your eyes. It is as simple and as complex—because the details of those specific mechanisms, in each case, will all differ enormously such that sometimes we will be able to study them but other times we won't—as this.

We have, in fact, already encountered an example of this notion of multiple mechanisms yielding the same result when we talked about epilepsy in preceding sections, because each epileptic patient may have a very different syndrome caused by completely distinct molecular and cellular events, yet the final result is identical in all of the patients, that is, the seizure: the abnormal high synchrony in the activity of millions of neurons that causes convulsions and loss of consciousness (sometimes without convulsions as in the case of absence seizures). So here we have this idea exemplified in a pathology: different causes, same result. And the advantage of this perspective is that if you want to stop seizures, you do not need to find out the particular molecular underlying causes of the seizures in this or that patient, you just need to try to stop the end result, the anomalous high neural synchrony, which can be done using medications like carbamazepine or even without chemicals in the case of neurostimulation of the brain using what is called deep brain stimulation (DBS for short) or transcranial direct current stimulation—the former is an invasive method, that is, electrodes need to be implanted in brain tissue, whereas the latter is non-invasive and is done from the scalp, and as a note for those interested in this topic of neurostimulation or "brain control", it is becoming very popular and applied not only to treat pathologies like epilepsy, Parkinson's disease and addiction, but also to things like improving memory or sleep, to such an extent that caution has to be advised because sometimes this brain control could go very, very wrong (Perez Velazquez & Estella Orero, 2020). These neurostimulation methods to stop highly synchronous neural activity have had great success in Parkinson's disease (where there is too abnormal high neural synchrony in certain brain regions) and a relatively good outcome in epilepsy, there being an enormous literature on the topic for those interested. Hence, we can appreciate the advantage of a high-level description of phenomena, at least to be applied to clinical settings.

In any event, to round up this somewhat long digression on equilibrium and non-equilibrium, just to emphasise that we need the latter to see interesting events happening in life particularly, because only under nonequilibrium conditions the state of a system can become unstable and a transition to an organised structure can occur. Such structures include oscillating spatiotemporal patterns in chemical reactions (you can watch the colourful patterns in the Belousov-Zhabotinsky reaction here[1]) and geometric arrangements of trillions of particles in other systems like in the Rayleigh-Bénard convection shown in Fig. 7.2 of Sect. 7.1.

And now that it is mentioned again, let us answer the question posed in that Sect. 7.1, why the Rayleigh-Bénard pattern consists of hexagons and not, say, circles or triangles. As it turns out, hexagonal patterns abound in nature, from the living to the non-living. They can be seen in the eyes of insects, in basalt formations, and of course, in beehives. The reason is that hexagon is the shape that best fills a plane with equal size units and leaves no wasted space. Try to use circles to fill up a surface and you will leave empty spaces. You can achieve a complete filling with rectangles or triangles... But, alas, the area of each rectangle when compared to its perimeter will not maximise the packing of the whole surface; only the hexagonal packing maximises area in relation to its perimeter. So, talking about bees, this means that each cell in the beehive has maximum space (surface) for the nourishment of the baby bee, and at the same time, the number of cells (progeny) that can be packed in the beehive is maximised. With squares, triangles or rectangles, the area for each cell may be large but there will be fewer cells, meaning fewer babies, and bees like to spread their progeny as much as they can. The words above have obviously anthropomorphised the tale, because each bee worker has no clue as to why it is doing what it does. Not that those bees understand geometry or mathematics, they just perform these feats driven by their genes, without the conscious knowledge we have. And in the case of the hexagons in the Bénard convection, the specific mechanism is that with the hexagonal structure the pull of surface tension in the liquid in each direction is most mechanically stable. Once again insisting on the high-level perspective of 'different causes same result', the essence of the phenomenon is the same regardless of molecular-mechanical forces in liquids or bees making babies: hexagons allow for that optimisation of the covering of a surface filled with interacting entities such that energy tends to be maximally exchanged, or dissipated in the terminology we have been using above.

It is hard to resist the temptation to include a quite technical note explanatory of this phenomenon of macro-order arising from micro-disorder, exemplified in the quote by Gnedenko and Kolmogorov: "In fact, all epistemological value of the theory of probability is based on this, that large-scale random phenomena, in their collective action, create strict non-random regularity" (Gnedenko & Kolmogorov,

[1] https://www.youtube.com/watch?v=jRQAndvF4sM.

1954). This apparently obscure and technical comment actually neatly summarises all we have discussed in the previous paragraphs. It means, in simple words, that if you put together a billion entities and allow them to interact and then examine the collective action of all those random interactions at the large-scale, macro-level, then you will find regularity, order, perhaps a nice pattern like that of the Rayleigh-Bénard phenomenon, other times a pattern not so nice like genocides in human societies. Because this applies not only to molecules but also to people living together in a society, interacting through various means that many times results in revolutions or atrocities; from this perspective, some historical events are easily understood. So now, you will know better when you experience the chanting madness of a crowd assembled in a football stadium cheering for the team to attack and score a goal or voicing in unison their discontent towards the referee of the match with the usual "florid" terms. While naturally the specific mechanisms leading to these collective phenomena vary depending on the system constituents and their means of interaction, it is somehow satisfying—at least for the writer of these words—that at this high level of description, so many natural phenomena ranging from chemistry to sociology have the same essence based upon the interactions and exchange of energy (or information if you wish) among the system's elements.

All this does not mean that if you just put together a massive amount of microscopic particles, be they atoms, molecules or cells (or macroscopic ones like animals or people), you will end up with an organised pattern. The recipe needs those additional elements aforementioned of energy dissipation, nonequilibrium/nonlinearity and a source of noise, and with those ingredients properly mixed, what at the micro level may look like messy, viewed at the macro level order, constancy, equilibrium, whatever you wish to name it, emerges.

We thus have started to see that complex behaviour emerges from simple rules. Further proof will be offered in the following section where in Fig. 8.1 diverse natural phenomena are displayed, all derived from the same basic rules.

8.1.2 Noisy Surroundings: When Noise Really Makes Sense

We need the third ingredient added to those of energy dissipation and nonequilibrium situations, to finally observe something that resembles life. Noise is needed. But not our everyday noise in our confusing cosmopolitan surroundings, the noise here is meant as fluctuations or variability in some components or characteristics of the system. In the case of our already well-known wonder of the Rayleigh-Bénard convection, the fluctuations are those occurring in the motion of oil molecules moving up and down and constantly interacting. In the example described above of the steel beads in a container separated in two chambers, some fluctuations that can be considered are the changes in the number of beads in both chambers of the container—recall that when one side starts to be more filled with the balls, it will continue to accumulate more little balls because they will lose more energy due to more frequent collision in this side of the vessel. There is as well the fluctuating velocities and motion of the beads, all these fluctuations generated by shaking the container, that is, by supplying energy to be dissipated.

Fluctuations, which can go by other names such as variability (although we would not use terms like randomness, for reasons to be explained in the following Sect. 8.1.3), are a most fundamental ingredient of our cosmos. In fact, according to some current theory, the whole universe comes from fluctuations. While we won't mention it here (it is rather specialised), those intrigued by the quantum world can find numerous writings about the universe being created from quantum fluctuations (fluctuations of what? We will leave it to those restless minds to try to understand these quasi-magical theories of the pre-Big Bang era). Nonetheless, it is true that we, humans and the rest of animate entities, come from fluctuations, and we do not need to go to the quantum level to see that. But before inspecting those fluctuations, let's try to clarify the significance of this noisy business for natural wonders to materialise.

Why are fluctuations of importance? Because dissipative structures amplify fluctuations (sorry again to use that technical term of 'dissipative system', but remember it is quite simple, it defines a system or structure that uses, consumes energy, exchanges it with the surroundings, and that's all you need to know for our purposes in this text). Normally, fluctuations are damped if the system is a closed one in a stable, equilibrium state, but there is always the chance that some fluctuations, helped perhaps by some energy input if the system becomes open, are amplified. These amplified fluctuations will incorporate more elements of that system such that from occupying a tiny space the fluctuation ends up occupying the whole system, at which point it is no more a fluctuation, rather a new state in the system has appeared. This is the path from being to becoming of fluctuations: they become (if amplified, of course) a possible new steady state in the system where they occur.

And this is nothing abstract or esoteric, we humans have witnessed this phenomenon all throughout history. Just think of a very stable and closed society where a few vociferous critical individuals are silenced—the fluctuation has been damped, thanks to the stability of the regime. Now shake a bit that society, add some financial trouble for instance, or social discomfort, open the country to others peoples and ideas—in the words of our current section, alter the equilibrium, push it towards non-equilibrium which can be easily done by going from a closed system to an open system—and then watch what happens. The words of those disapproving individuals may not be easily damped, rather become amplified, driving the system (the society) to a new state, and we have the beginning of a possible revolution. It is crucial to realise that the success of this new state raised by the amplification of the number of unhappy individuals creating those fluctuations depends on what is out there, on the environment, what in science is known as boundary conditions. So if this society is surrounded by others that are powerful and that do not care too much about revolting, they may just quench the uprising in their neighbours by force. This simple example illustrates the wonder of the creation of new states by amplifying noise—and we mean not only states as countries or societies, but conditions in any natural system.

Thus, this is the significance of variability/fluctuations in the dissipative system, the generation of new states, new conditions and new patterns. Human societies and the rest of living systems are dissipative structures because they are maintained by flows of energy (and matter, we all eat, don't we?), therefore, it should not be too surprising to see organised patterns of activity in living phenomena that may become more and more complex. Not only in the living but also in the non-living, because most of these inanimate inorganic systems (living systems are organic, that is, based on carbon atoms) are also dissipative, as we saw in the steel beads example. Should it then be expected to see similar patterns in the lifeless and in the biological? Please check out Fig. 8.1 and, without looking at the figure legend, try to distinguish which ones belong to living systems and those that are non-living—and those who think in terms of an unambiguous and definite separation between the animate and the inanimate are especially invited to consider the shapes shown in that figure. We have there neuronal axons and dendrites, lighting during a thunderstorm, river networks, branches of trees and lung bronchioles. The task is not that easy, eh? Anyhow, this is not the place to answer the perennial question that has occupied minds since the beginning of cultures about what differentiates the living from the non-living (but in Sect. 9.1 you will see some words addressing this, most fuzzy border between the organic and the inorganic).

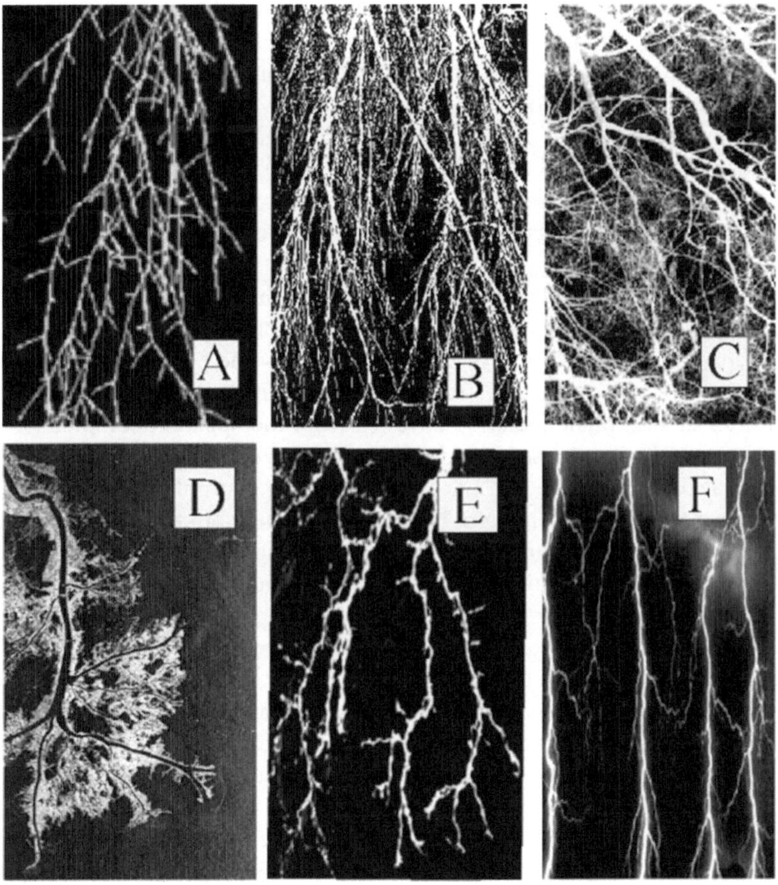

Fig. 8.1 Dead or alive? Common structures in natural phenomena. (**a**) dendrites of neurons; (**b**) tree branches; (**c**) lung bronchioles; (**d**) river network; (**e**) axons of neurons; (**f**) lightning. (Adapted from Perez Velazquez, 2009)

Now you can look at the figure legend. This exercise has hopefully served for readers to note the striking common (structural) aspects of the natural phenomena displayed there. This is just a tiny sample, for these tree-like patterns are ubiquitous in nature from mountain ranges to snowflakes and metal-salt sediments. Hence, are we, living entities, so special after all? We are, but perhaps not so much as some may think, please go on reading the rest of this Part II to comprehend our authentic speciality. A fine start to transcend our created arbitrary dichotomies is to take a deep look into the differentiation between the animate and the inanimate, and in so doing, searching beyond dichotomies and from a global perspective on the natural laws, a clear partition of matter into animate and inanimate becomes ambiguous, as readers have probably just experienced trying to classify the panels in that figure.

It would be unfair not to mention that some cultural traditions seem to understand the hazy borders between the animate and the inanimate; they have more holistic perspectives. One example is that of the North American natives, who have advocated for legally declaring a river a person. The Magpie River in Quebec (Canada) has been granted legal personhood, so now it has rights. To consider a river a living entity is not inconceivable, because while the water and sand may be "dead", the fish and microorganisms and other animals living in it are "alive". Concepts of environmental personhood have been developed in other countries like New Zealand, where a national park became a "legal entity". These ideas about environmental personhood can be inspected in some papers (e.g., Gordon, 2019). By the way, while these stories may seem remarkable, the fact that corporations are granted personhood too is taken as normal—this is called *corporate personhood,* the legal notion that a corporation, separately from its associated human beings, has some of the legal rights and responsibilities enjoyed by natural persons.

A natural question to ask is whether these common shapes in very distinct phenomena arose as a coincidence, or whether there is some principle that guides the development of these patterns of matter. Can a simple universal logic ruling the evolution of natural phenomena be unveiled from the apparent complexity of the natural world? The specialised paper addressing this query appeared in 2009 with the main purpose, expressed in the title, of "Finding simplicity in Complexity" (Perez Velazquez, 2009). What are, in very simple and non-technical words, those general principles that presumably brought us, and the rest of biological entities, to life? It turns out that fluctuations or noise is a fundamental ingredient in those principles as much as it is in our recipe for interesting things to occur.

In simple words, the fundamentals of those patterns in Fig. 8.1 are these three: competition, cooperation and fluctuations—the very basic foundations of pattern formation. Readers familiar with Darwin's theory of evolution may have connected these three elements with the theory, because his ideas can be summarised as the selection of the variation through reproduction, which can be rephrased in terms of competition (selection), cooperation (reproduction) and fluctuations (variation); competition and cooperation among fluctuating sources lead to the selection or transient stabilisation of structures. In ecosystems, where Darwin's thoughts apply, these structures are evolutionary patterns of biological entities, but also could be the structural ones in Fig. 8.1, or patterns of cellular activity in organs of the body. To wit, it is known the importance of competition and cooperation between brain neural networks that result in our mental lives and actions—cooperation among neurons may take the form of synchronisation of their activity, and synchronised neural networks compete to gain "control" of the next stage in the chain of neural activity, recall from Sect. 1.4 those never-ending chains of neuronal communication.

These close relatives, competition and cooperation, are ubiquitously found in natural phenomena and represent a fundamental principle of organisation. As a small sample of instances beyond the nervous system, consider the competition and cooperation between biochemical and mechanical process in phyllotaxis (the arrangement of leaves on a plant's surface) or the competition between external forces (gravity and wind) in the formation of branches and twigs in the structure of trees; the snowflake patterns are also determined by the cooperation of water molecules that tend to unite together and the competition between the diffusion of these molecules and properties of the solid-air interface like surface tension; and finally, how about the "chemical gardens", tree-like structures shaped by the competition between internal pressure and membrane formation, which grow when soluble metal salts are combined with solutions containing silicates. Naturally, our cultures and societies are also clear examples of individuals cooperating and competing, something that has been with us—humans and other animals alike—since the beginnings of time (hopefully it has become already apparent to readers how we are jumping levels of description when describing most of the principles and ideas mentioned in this part II, from the microworld of particles and molecules to macroworlds of groups of beings and cultures, a sign that same principles explain natural phenomena at all scales… We may not be molecules, but we seem to be governed by similar rules).

Hence, from phylogenetic diversity to the formation of snowflakes or to mental activity, competition and cooperation lead to the selection of fluctuating patterns that become the new states of the system, be it an ecosystem, a geological formation or a nervous system. This is all we will say here regarding these notions lest it becomes too specialised, but as always, we will recommend some basic writings for those interested, like Sonya Bahar's "*The Essential Tension –Competition, Cooperation and Multilevel Selection in Evolution*" (Springer, 2018), or the aforementioned article "Finding simplicity in complexity: general principles of biological and nonbiological organisation" where all these thoughts are explained in great detail. Now, a brief technical note for those with holistic mindset tendencies: a most basic, and quite simple, physics equation widely used to describe natural phenomena represents all these ideas described in this section, it is the Langevin equation (after Paul Langevin, 1872–1946), composed of a deterministic term (which in principle comprises cooperation and competition because these cooperative or competitive interactions are regulated by deterministic laws ruling the particular entities forming the system) and a stochastic (noisy) term representing fluctuations, so here are encapsulated in a mathematical formula a good extent of the principles that build natural phenomena.

8.1.3 A Fluctuating World

After all the previous sections that sounded like a eulogy, praise of variability and fluctuations, it is time to be a bit more specific about what fluctuations may have driven life to emerge and us to be here. But before, let us be precise as well about something that may be misleading. To some, saying that fluctuations, in whatever property or particle, have played a fundamental role in the emergence of life may sound as if life appeared because of randomness, since popularly randomness is associated with the variability given by fluctuating events and things in general. As a matter of fact, the notion of randomness is ever-present in our everyday life, where close relatives like luck, chance, fate and so on are constantly invoked by people in all sorts of situations; randomness is so much used and abused. But—and perhaps this comes as no surprise anymore—randomness is a very relative concept, yes, very much another of our conceptual constructions like those aforementioned of probability (second paragraph at the start of part II), consciousness, life, equilibrium, you name it. So, considering how many times we invoke this notion in our everyday situations, let us allow for a very brief explanatory digression.

Let us note that the notion of randomness is relative and depends on what information we already possess. It is very easy to see this. We shall use a tiny bit of math to demonstrate this point (nothing to fear, for after all mathematics is the art of counting, and everybody knows how to count). The number *e*, which is the base of the natural logarithm, is a constant number used a lot in science whose digits after the decimal point appear random (they start like this 2.718281828459045… and so on to infinity because the sequence never ends). The truth is that this sequence may be random or deterministic depending on how it is "measured". Statistical tests applied to the sequence of digits will indicate randomness since the sequence cannot be predicted at all. However, we do know how to perfectly generate those digits using a nice compact formula, namely $e = \lim_{n\to\infty}\left(1+\frac{1}{n}\right)^{n.}$ and this is not the only formula that can be used to generate all those apparently random digits; there are several more, so, here we have deterministic equations from where we can obtain the "random" sequence of decimals. Hence, we must conclude that it is not random but totally determined by the mathematical expressions. The same thing occurs with another extremely important number in science, number pi (π) which is 3.141592653589… going on forever without any repeating pattern in the sequence of digits, so if you apply any statistical test to the sequence, it will tell you the series cannot be predicted hence it is random. Yet anybody, any child (who knows division!) can calculate it, because, as we learnt in school, pi is the ratio of the circumference divided by the diameter of the circle. So draw a circle and measure the perimeter and the diameter with great accuracy and behold, you have a good approximation to pi; the better the accuracy of your measurements, the more decimal digits you will obtain. And also, like in the case of number *e*, there are very neat formulae from where you can obtain that sequence, for instance, $\pi = \int dx/(1 + x^2)$, so here it is again a deterministic equation from where to obtain that apparently "random" series of decimals. Then, are numbers e and π random or deterministic? It depends on the viewpoint, depends on what you know. If you do not know math and only know how to apply statistical tests, the numbers will be random to you, but if you are proficient in mathematics, you will know the precise deterministic equations to generate both numbers. What does this mean for our lives? That we can never be sure something is really random, because it may be that we just do not have the knowledge to see its regularity.

And what does this say about those fluctuations, which is the reason we talked about this anyway? That the variability, the fluctuations, may be completely determined by deterministic laws, but because we cannot keep track of all that myriad of things and events occurring, we will never find out the end result and will not be able to predict accurately the future. As an illustration, recall that it was said above, at the start of Sect. 8.1.2, that fluctuations in the velocity and position of the steel beads in the shaken chambers result in changing their number in the two sides that will determine which one fills up and which becomes empty, but the motion of these particles is determined by physical laws we know very well, the Newtonian dynamics—that most classical physics which children start to learn in school—, hence, these fluctuations in their movements may be completely deterministic but nonetheless we cannot keep track of each individual bead and calculate how it will move, where it will go, what velocity it will reach and consequently, to us, the whole thing looks like a random mess, an example of stochasticity.

Thus, it may be fairer to use a term that does not really imply a non-deterministic world, rather denotes our ignorance about what could be totally deterministic but that we cannot predict, the term is stochasticity. It is defined as the quality of lacking any predictable order or plan, very much like randomness, but this does not mean the underlying laws of the particular stochastic phenomenon are not deterministic, it only means, like in the case above of the steel balls, that we are ignorant about—we cannot follow or compute—each possible individual event. Stochastic fluctuations are those we are talking about, which may be due to precise and deterministic laws but since we cannot monitor all the processes, for us they look like chance, random. We have obviously greatly simplified these musings on stochasticity and randomness, the considerations could become quiet technical as these are very commonly employed concepts in several branches of science. From an intuitive viewpoint, the one we are interested in a book like this one aimed at laypeople, both terms tend to be used as synonyms, but one—stochasticity—implies certain determinism, while the other—randomness—means complete non-determinism, a lack of any principle of organisation. In any event, whether truly random or stochastic, the final result is unpredictability, so perchance a better word to use, rather than random or stochastic phenomena, would be *unpredictable* phenomena. But still, the same limitations as described above about the decimal digits apply here: whether you can predict it or not depends on your knowledge, so it seems impossible to remove the ambiguity in the concept regardless of the term used. One wonders whether Einstein's words that "space and time are modes by which we think, and not conditions in which we live" could be rephrased in terms of these other concepts as "randomness and determinism are modes by which we think, and not conditions in which we live".

Randomness then joins other scientific concepts, some very fundamental ones, that have been shown in this text to be somewhat ambiguous, relative. At this point, in some readers' minds, a feeling of hesitation in science may have been aroused because after all, wasn't science supposed to be built using unambiguous, precise and definite notions? Rest easy, all is fine, we scientists, at least the majority, realise these limitations of our intellects and our technology, the Nobel Prize-winning biochemist Christian de Duve being a good illustration of this when he said that "We must accept our concepts for what they are, provisional approximations that are as much fictions of our minds as they are faithful depictions of the facts". Trouble arises when prudence gives way to belief and subsequent strong attachment to our ideas, which applies not only in the realm of science but also in many domains of human affairs. If we devote too much attention to our artificially/arbitrarily created concepts and then we start to believe blindly in these notions and in their strict definitions, then we run into trouble. These concepts help us talk and communicate ideas, but we have to take them lightly, for these are our man-made inventions and should not be objects for the belief in eternal truths.

Perhaps some apology is in order for the previous extensive foray on fluctuations, equilibrium, energy dissipation and related contemplations, but these are important things to understand how patterns are formed in nature, because life is, in the final analysis, a pattern of organised activity. In the words of E. J. Ambrose, "The matter in life has no permanence; only the pattern according to which it is arranged or organised has permanence. Life is basically a pattern of organised activity" (Ambrose, 1982). The quote, that appeared in his '*The Nature and Origin of the Biological World*', captures the title of our book, that matter is in constant being and becoming, although even the organised patterns are also mutating; perhaps some inherent laws that govern the formation of the patterns may be permanent but certainly patterns change too and become something else. And what could then be the essence of those perchance immutable physical laws underlying the mutating patterns found in nature? We have already seen three most conspicuous features in our recipe to cook interesting things: energy dissipation, nonequilibrium conditions and variability—fluctuations, noise, whatever you wish to call it (and please sprinkle some nonlinearity too, we decided not to mention this point because it may become too technical but nonlinearities are also essential for these phenomena to develop, in fact almost all natural phenomena are nonlinear, which in crude words means the output is not the straight sum of the input, like when you get anxious or not about something depending on the situation—and this could be called nonlinear psychology!). But as to whether these features are permanent or not, well, let us only mention that there are those who think that even the physical, natural laws, does not matter how primordial and fundamental these are, are not immutable. One of these characters, the theoretical physicist John Archibald Wheeler said, "Not law but mutability has the last word" (Wheeler, 1977). If you read the article where he wrote this, you may find out that all, even the most basic laws, can become something else, or, in his own words, be transcended (the suggestive title of his essay is '*Genesis and Observership*' where he discusses the importance of us, observers, in the genesis of reality). Nevertheless, so as not to become too involved in heavy philosophy and theoretical physics, let us continue working with those three ingredients for the recipe for interesting things to happen and put these together to cook life.

What fluctuations can be contemplated that pushed us into existence? As a preliminary exercise, let us consider possible fluctuations that bring about some patterns portrayed in Fig. 8.1. For example, in the riverbeds, variations in the composition of the soil—whether rocky terrain or softer materials—, variability in rock heterogeneity, all these fluctuations in the earth through which the water flows determine where it will flow, which in turn will change the soil properties and create more variability, the end result will be the river bed/basin. In the example of lightning bolts, the fluctuations are those present in the air, because a lightning bolt is very much an electric current moving between clouds and earth (or between clouds), and the energy released—or dissipated, because, yes, bolts are dissipative structures too—is seen as light (and heard as a thunder because of the expansion of the air due to the extremely high temperatures occurring in that path of electricity) and the path taken will depend on the fluctuations in the composition of the air, especially ionised particles that give air its conductivity to electricity; as it happens this electric current takes the celebrated "path of least resistance", travelling where the air conductivity is greatest so that it can reach the ground with the optimal path, and it is jagged because, although it looks like the bolt is formed all at once, it is actually produced in many steps and each one takes the "easiest" direction. So now you know, in a global sense, why lightning may strike you but not your neighbour.

And recollecting phenomena described in past sections, the network build-up by the slime mould we met in Sect. 1.3 to transport nutrients between distant parts of its body depends on fluctuations too; recall the motility pattern the mould made while foraging for food (shown in Fig. 1.5) constrained and determined by the varying places where the nourishment was found and the distances between them. Your brainpower also springs from fluctuations; if you look at your neurons you will see those lightning-like structures of dendrites and axons (panels A and E in Fig. 8.1) and as we explained in previous sections, (1.4 and others) these constitute the structures through which neurons communicate. The neuroscientist Santiago Ramon y Cajal already observed in the early twentieth century that axons without environmental interferences grow straight, so adding some varying interferences, coupled to chemical reactions occurring inside and outside the cell, result in axonal growth and guidance by chemotaxis and determine where the axon happens to make contact with another cell and establishes a synapse. So, thanks to all those fluctuating 'interferences' that your neurons encounter while finding their way into your brain, they have complex axonal and dendritic structures that allow them fruitful communication. This structural variability in the geometry of neural connections finds its dynamic counterpart in the so much talked about importance of the variability in neural synchrony patterns of activity we saw in many previous sections, such that fluctuations in neuronal synchrony are associated with good brain health whereas lower variability is found in pathologies like epilepsy or coma.

We see then that fluctuations are not only crucial to forming structures, but are a factor dictating dynamic patterns in nature. Hence, changing now levels of description (something that is always fun) and leaving the structural level, at the conceptual level we find that the spread of ideas in human societies is also constrained by variability, the fluctuations in the individuals' mindsets; and in ancient times, when people could not travel too easily due to archaic means of transport, the spread was limited as well by geographic barriers. Thus, from structures to cultures, variability is an essential ingredient in the formation of patterns.

An entertaining way to see the essential contribution of fluctuations in the making of our world is the example of fractals. Many readers undoubtedly have heard and read about fractals, structures that are complex self-similar patterns across different scales. The rule to create them is simple, by repeating a simple process over and over in an ongoing feedback loop. But, alas, those patterns do not resemble too much the natural ones, perhaps some do resemble coastlines, but not much more. Now, add to that code that produces the fractal a random number generator and see what emerges (note: these are not really random numbers as these are generated by computer codes, hence they are deterministic, remember the above-discussed considerations on randomness). This procedure simply adds fluctuations in the application of the main rule in the computer code generating the fractal structure, thus they are known as random fractals—insisting, they are not really random!

Figure 8.2 shows two typical fractals in the row above, and two random fractals below, the latter looking very much like natural structures (the one on the left looks like the kidney network of capillaries from this angle of view, and changing the angle will look like broccoli, as you can watch here[2]). Random fractals are popular today when it comes to modelling clouds, landscapes and other natural phenomena for the purpose of computer image synthesis. Fractal structures, by the way, abound in our bodies—neural networks and alveoli in the lungs are two examples—as they do in the rest of the natural world. This is not an accident nor a coincidence, they are ubiquitous because such structures are very resilient to breakdown or dysfunction. Loss of the fractal structure in physiological systems is the subject of current research that shows that loss is associated with pathology.

To learn about the variability that gave rise to our living forms, the very essentials of life have to be considered. All living material depends on (bio)chemical reactions, which is called metabolism. What we term life originated in a primordial soup, so to speak, the oceans of ancient Earth which contained a very large number of chemical compounds floating around and interacting, those molecules that met and were able to exchange energy by means of a chemical reaction would react, those that were not able to "chemically communicate" would not react. That we are the ocean's offspring can still be seen in the composition of our blood, as the plasma (the watery part of the blood) has a similar salt composition to that of the sea—sodium, chloride, calcium, magnesium, etc.—although the concentration of each

[2] https://www.youtube.com/watch?v=ydp0ogLg5Qs.

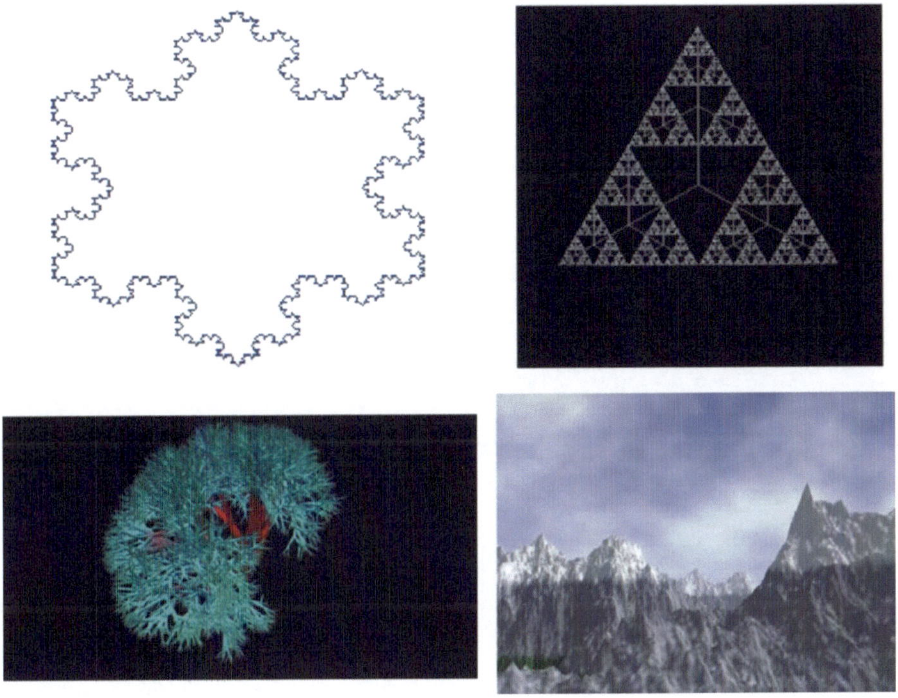

Fig. 8.2 Standard fractals (above) and random fractals (below): the beauty of adding noise

salt differs from that found in seawater nevertheless is a good hint about where it all originated. The same can be said of another important fluid in our bodies, the cerebrospinal fluid present in the nervous system, although because it is derived from the plasma the similarity with seawater is therefore expected. And as usual, we offer a brief technical note: plasma contains molecules like proteins, fats and other materials that are not so common in seawater, but this is natural because proteins and lipids are synthesised and degraded constantly in our bodies, this being part of our metabolism.

The primordial variability in those primitive waters that were essential for what we call life to start emerging was provided by, among other factors, fluctuations in the composition of molecular compounds that gave rise to a great number of possible chemical reactions, from which an early metabolism appeared. Afterwards, once this early (bio)chemistry was encapsulated into cells, other crucial fluctuations that moved lifeforms ahead were those in genetic structures (called mutations). But let us not jump ahead. Now, equipped with the recipe for interesting things to emerge, we are ready to create life.

Chapter 9
Let There Be Life

Contents

The basic principles of the formation of structures in nature, or patterns both structural and dynamic, have been outlined in Chap. 8. We started seeing how thanks to the three main ingredients for interesting things to emerge, what seems improbable is many times very probable. Of special importance for this to happen is that many interacting elements are put together in a place where there are energy sources, and those components are allowed to display variability in their means of interactions. Does it mean if you combine a bunch of chemical compounds in a sea and sit in front of it for a million years, you will find life? Well, it depends on where you do it. If you do it in Jupiter, the molecules will soon freeze, and besides, its ocean is made not of water but of liquid hydrogen, so your chemicals will not stand a chance to combine into anything meaningful, at least according to what we are now pursuing. But if you are on a planet similar to Earth, your chances to see features that form part of what we term life is almost guaranteed. Just like the emergence of the sense of personal identity is inevitable as we saw in Part I, now, we will see how life is inevitable too on a planet like ours.

What follows is necessarily a very summarised and simplified vision of modern thoughts about the origins of life. Whereas these ideas remain theories, what will be presented here is the most accepted scenario among scientists. And yet, how the earth was seeded with the building blocks of life is the topic of controversy too, from the premise that the primordial elements were present on the Earth from its origin to the theory of "panspermia", where, as in mammalian reproduction, the

J. L. Perez Velazquez, V. Nenadovic, *Being and becoming*,
https://doi.org/10.1007/978-3-030-78264-1_9

ovum of the earth was "seeded" by elements from comets and other space nomads. In any event, a theory can never be proven, only disproven, and the case is worse with theories about the origin of life because we would need to go back to those early ages, so until someone invents time travel, a definite clearest indication that these proposals are right will have to wait. Having clarified this point about theories, let us explore the most possible scenario that developed long, long ago.

The following is what very likely occurred.

9.1 Molecular Crowding: A Tale of the Most Probable

Something that has to be clarified is that some precise details of these process that will be described (in as simple words as possible) are still unknown, but the global scenario is inferred from a great number of scientific observations, experiments, and thoughts of a multitude of scholars in various disciplines. And note that there was not a unique event that sparked life; the prevailing view is that the transition from non-living to living entities was not a single event, but a process that took a long time involving features like self-replication and self-assembly of molecules, autocatalysis, and the emergence of cell membranes. Because time is another essential ingredient of the recipe to prepare life, the happenings detailed below had plenty of time to develop. If you want to rush and do something in a very short time—one week, to be precise—the most you can do is to create some molecules found in today's organisms. This was actually achieved in 1952 by a graduate student and his supervisor, Stanley Miller and Harold Urey, who performed a chemical experiment—baptised after the creators, the Miller-Urey experiment.

The experiment simulated the conditions thought to be present on early Earth, for which they mixed in a flask water and a few gases believed to be abundant in those times (oxygen was not included, this most fundamental gas for us appeared later on, courtesy of cyanobacteria and other microbes), and electrical sparks were fired between electrodes within the flask to simulate lightning. After just one week the reaction was stopped and, analysing the created materials, they found many molecules that form part of us today, like amino acids. This experiment inspired several others that added more evidence that biomolecules can be generated in the conditions that are thought to be present in the early epochs of our planet. But naturally, this is not life as we know it. To obtain a more efficient pattern of chemical reactions (an organised pattern, a connected metabolism, how this occurred is detailed a few paragraphs below), a long time is needed, because to become "alive" these molecules have to do something else: to react, to interact, to combine, to create new molecules and cells and organisms, entities that are one thing and become another, so molecules grouped in cells, and cells formed organs and creatures that evolved one way or another, some changes were successful and others not so (and we mostly see the successful ones, the others became extinct as nature does not give too many opportunities), and all that takes a long, long time.

And to give a perception on time, here are some "time facts" for your information and perspective. The Earth formed approximately 4.5 billion years ago —or as some Europeans would say, four thousand five hundred million years, this may not be a googol (that we met in Sect. 8.1), but it is an imposing number, 4 followed by 9 digits, 4.5×10^9 years; and life probably began between 3. 5 and 4 billion years ago—the earliest evidence of life found so far comes from fossils discovered in Australia that date back to about 3.5 billion years ago (these prehistoric times are estimated based on the decay of radioactive isotopes). And if you are curious about this earliest of all living creatures discovered up to now, they are cyanobacteria that binding together formed the found fossils in structures known as stromatolites, and following the principle that "if something works keep doing it", present-day microbes make stromatolites too.

This text is not the place to go in depth about the major theories of the origin of life, where we could discuss the Oparin-Haldane hypothesis or the RNA world hypothesis and of course the aforementioned Miller-Urey experiment. Rather, we will focus on the essence of what may have occurred. Nonetheless, the Miller experiment and others that followed showed that biomolecules could be generated from inorganic materials in the early Earth environment, so let us take this as the starting point and inspect what could occur once these compounds that today form part of our bodies appeared in the arena of early Earth chemistry. And if you think these biomolecules are very special and unique, then think twice: many of our most important chemical constituents—bases of the DNA that constitute our genes, amino acids, etc.—have been found in meteorites that fell on Earth from outer space, and this is an early sign that we are not as special as we may think, that this life business may not be as improbable as we imagine... Proceed reading, please.

The evidence from the previously mentioned experiments where biomolecules were sort of spontaneously synthesised suggests that life—that is, organic molecules needed for life—arose gradually from inorganic compounds which formed what we call today organic materials, those that have carbon atoms in their structure. Some may think that because some of these molecules have been found in meteorites, these may have come from outer space. As said above, for our purpose here, we do not care about fine details; whether those compounds originated in the early prebiotic chemistry of Earth, whether they came riding meteorites or were planted by aliens, our mission is to inspect what happened next; and in any case, postulating that these biomolecules were brought by extraterrestrials does not solve our query about the origin of life and what this is all about; it just places it on another planet.

Thus, let us try to follow what events developed due to the great number of these chemicals on our planet, and especially in the sea, where many of the reactions that in the end formed a metabolism took place. Water is best suited than solid terrain for molecules to react, and the primaeval sea where molecules were happily interacting is known as the *primordial soup* (indeed, we are cooking life). What occurred, then, in very few words, is that the development of a set of chemical reactions reached at some point an organised complexity such that characteristics associated with what is called "life" (self-replication, compartmentalisation, and so on) emerged: a primitive metabolism arose in that primordial soup.

Why did this sort of organisation occur spontaneously? To some extent, we have an answer from what we have seen so far; the three main ingredients for our recipe were present in these early times: energy was abundant, taken, and exchanged (that is, dissipated) in many chemical reactions, and a huge number of fluctuating molecules were floating around in a system that was not at perfect equilibrium. But why is life based on organic compounds? What is so peculiar about organic chemistry?

Here, one should recall the comments on equilibrium in Sect. 8.1.1. It was there mentioned how the processes we find in nature are almost never at equilibrium, rather in a sort of unstable state far, or sometimes not so far, from equilibrium. In reality, what is observed in nature is *the tendency to approach the most probable state*. This is perhaps the best manner to think about that intricate issue of equilibrium, that is, processes always try to attain their most probable state—this is equilibrium—but because they are open systems exchanging energy and matter with surroundings, this state is hardly reached, and if it is ever achieved it will be displaced soon by energy inputs or fluctuations in the components; nonetheless, the tendency remains, and systems strive (anthropomorphising again, because there is no struggle, no purpose, it is spontaneous, it just occurs) to approach equilibrium only because it is the most probable state.

What does this mean in the case of a planet like ours with a variety of molecules floating around? The most probable state would be reached by the maximisation of the number of chemical reactions. It is like when you have a bucket full of balls and throw them onto the floor of your room, there will be many collisions. It would be miraculous if only 2 or 3 balls collided; most likely all balls will suffer a collision with others. Imagine now each ball is a molecule and each collision a chemical reaction: the most probable state is that in which many molecules react with others, provided of course they are chemically compatible—while this is not a chemistry course, let us just specify this most crucial point that only molecules "chemically compatible" (that is, they can share electrons or energy) can react and, among other things, form ionic or covalent bonds creating new molecules. These two are fundamental chemical bonds present in our molecules, but not all reactants will be able to materialise them, so a better balls-in-a-bucket simile would be to carve different keyholes in some balls in that bucket and have other balls with protruding keys that fit specific holes, only those balls that have complementary key and keyhole can "react" upon colliding on the floor. In any case, the most likely result is that of a large number of collisions/reactions.

So why a carbon-based life? The previous paragraph on reaching states with many chemical reactions indicated the reason. In those primaeval ages, our planet had lots of carbon and molecules with carbon and hydrogen, called organic compounds, along with inorganic compounds. Our body, life as we know it, is organic: a bunch of molecules containing carbon atoms. What is so peculiar about organic molecules? You can find the answer in a library, but not by reading the books, rather by looking at them! If you have ever taken a look at textbooks on organic chemistry, you may have noticed these tend to be thicker than those devoted to inorganic chemistry, and this is because there are more possibilities for organic reactions; in fact, the number of organic compounds is much greater than inorganic compounds due to the special ability of carbon atoms to join with other carbon atoms in chains, rings, and various geometric arrangements (mind you, there is also organometallic chemistry which is like the combination of organic and inorganic chemistry, but let's not get into complications now). So the most probable state for ancient Earth included the maximisation of the number of chemical reactions, and thus, carbon—or organic—chemistry rapidly became prevalent due to their sheer number of possibilities for reactions, and soon this chemistry, or a particular set of reactions within that chemistry, resulted into biochemistry: the origins of what we term life.

Those versed in chemistry may know that silicon (Si) has similar chemical properties as carbon (belongs to the same group in the periodic table of elements) and it is very abundant too, more than carbon in fact. Can a silicon-based life be conceivable? Perhaps, but not here, because the properties of carbon prevail within typical environments at Earth's surface; without going into details, due to its peculiar chemical properties provided by its electrons, carbon bonds to other atoms more strongly and particularly to itself, forming chains that are present in our molecules and polymers (polymers are materials made of chains of molecules, like proteins and DNA). And to make matters worse for silicon, in a water-rich environment like ours, silicon's chemical capacity is limited because of silica formation (the major constituent of sand). Nevertheless, in another planet under other situations, silicon may possibly be the basis for an alternative biochemistry. The point to be understood is that as long as the conditions allow certain elements to react in complex manners with many possibilities of reactions, other biochemistries are possible. It is a matter of what is more probable—something that is already appearing to be a recurrent theme—, of what conditions allow for a greater number of configurations of molecules and chemical reactions, because in the same manner that you cannot think too much with only a dozen of neurons, there cannot be much of life with a dozen of molecules and chemical reactions (does not matter in how many ways they react among themselves. they are just too few). Hence, what is special about our planet is that its components and conditions permit carbon and its associated organic chemistry to flourish —that's all. In other worlds there could be a biology quite different from that on Earth. So next time you are on a sandy beach consider that the major component of that sand, silicon, may be part of extraterrestrial life forms (note: this word can be also spelled as life-forms or lifeforms; in this text, all will be used).

Having established that organic chemistry has enormous possibilities for a multitude of chemical reactions, the next inquiry is what could have happened in that primordial soup where all molecules were reacting will all (again, depending on chemical compatibility). Under these conditions, sooner or later, some molecules will become entangled into a web of reactions where one synthesises or catalyses the synthesis of another, in a sort of loop where feedback and feed-forward reactions are present. In scientific words, a *connected metabolism* took place, or perhaps at that early stage instead of metabolism we could call it a connected web of chemical reactions. You only have to open a biochemistry textbook and take a look at figures where biochemical reactions are depicted. What you will see in many of

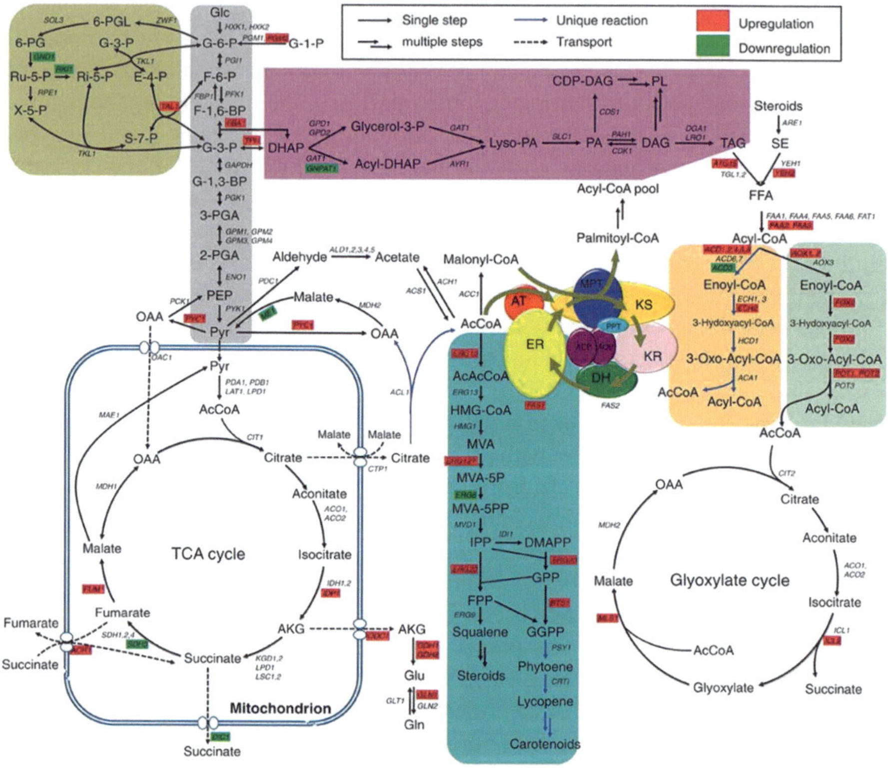

Fig. 9.1 An illustration of a connected metabolism. Biochemical materials being synthesised and destroyed in chains of biochemical reactions, reaction products acting on other reactions, feedback loops and catalytic mechanisms, virtually every molecule being created by another molecule in this biochemistry pathway (please ignore details!)

them is something like the example shown in Fig. 9.1 representing lipid biosynthesis pathways (taken from Zhu et al., 2012): loops, cycles, and molecules created and metabolised, whose products act on other reactions—sometimes as catalysts, meaning that the compound facilitates a specific reaction—in feedback or feedforward manners. If you put yourself in the place of an atom in a molecule, you will see that you can travel from molecule to molecule, being exchanged in various chemical reactions, and reaching almost any other molecule in the biochemical pathway. Indeed, cycles are fundamental biochemical pathways; to wit, the urea cycle (thanks to which we get rid of toxic ammonia and excrete it as urea in the urine) and the Krebs cycle, that goes under other names such as the citric acid cycle or TCA cycle (which allows organisms to release energy stored in molecules to be used in all sorts of biochemical mechanisms), are two prominent examples, but there are a myriad of these loops running constantly in your body.

How from a set of reacting molecules—what could be considered an early metabolism—features associated with life appeared has been partly answered, at least from a global perspective, in the works of Freeman Dyson, Stuart Kaufmann, and others. This high-level view could well be the most informative level to consider these questions, at least to obtain a general understanding, but naturally, the specific, lower-level mechanisms and processes can be investigated as well. In straightforward words, the basic idea is that life evolved from very complex chemical sets of reactions that crystallised in other simpler. In a bit more sophisticated words, when systems with varied chemical reactions reach certain complexity — meaning with lots of reactants and products of reactions that interact in a great number of ways to continue producing chemical transformations—, autocatalytic sets emerge spontaneously and catalytic closure is achieved in the sense that each polymer has one step in their formation catalysed by some other polymer (recall that a catalyst is a compound that facilitates a specific reaction). In short, a sort of a wide web of interrelations among all of molecular pathways in that system emerges. Stuart Kauffman originally proposed this idea of "catalytic closure" when chemicals started to auto-catalyze themselves forming a "closed' web of chemical reactions endowed with certain complexity. The advantage for this phenomenon to occur in a set of chemicals is, in uncomplicated words, that a self-sustaining network of chemical reactions has emerged such that the components are produced (synthesised) by themselves, by components of the system itself; a completely closed set would not need to look for other components outside to make more molecules of that network.

One brief digression. We are talking about life, about when features associated with life emerged, and here we encounter the same predicament as in Chap. 7 on the definition of consciousness, in that it seems impossible to define life. After centuries of attempts, there is no globally accepted definition of life. Is a virus alive? A bacteria? There is no need to repeat the reasons for the advice given in that chapter about the advantages to disregard specific definitions of life and consciousness, and instead use characteristics to describe them. Otherwise, J. Swain's counsel should be followed and develop "many definitions of life some more useful than others depending on the situation in question" (Swain, 2002). But if we choose the enumeration of features, then life can be defined by characteristics like self-reproduction, exchanges of energy, genetic transmission, compartmentalisation, etc. And like in the case of consciousness, the scientific study of life becomes more defined as each of those features can be investigated separately. So if one is studying genetic materials he is studying life, if another is examining the thermodynamics of lipid assembly, she is studying life too. And just like in the consciousness business we concluded that entities like microbes possess a few very basic features of consciousness and thus can be endowed with some partial cognition, same occurs here in the case of the living and non-living: it turns out that some inorganic materials have some of these properties typical of the living and could be considered at least "half alive" —how about self-reproducing inorganic colloidosomes, or "protocells" made up of chemical reaction networks designed to model pre-cellular systems. It is the same old story, these are our man-made notions that try to unambiguously specify phenomena but out there in nature there is a continuum and trying to impose clear demarcations may lead to misinterpretations; remember Fig. 8.1, similar principles of organisation apply to the living and the non-living.

Back to connected chemical reactions. Naturally, these theories about connected sets of chemical transformations are full of details about mechanisms which may have produced such autocatalytic systems of reactions, and we could talk about how large the set of polymers that can act as substrates and products of reactions should be, how many reactions should take place, and a host of other details that are beyond the scope of our text. All these variables and parameters have been used in mathematical models attempting to find out how easy or difficult it is for these connected, or autocatalytic, sets of chemical reactions to appear in the scene of the primordial soup. Those curious minds can find enough food for thought on these topics in Dyson (1999) and Kauffman (1993).

In reality, there is no need to place great emphasis on that concept of catalytic closure—it is fair to mention that some argue against this phenomenon being the crucial one in the development of the metabolism, we scientists like debates!— because in the final analysis, the fundamental idea, the take-home message if you will, is that a set of reactions became localised in a small volume separated from the environment through a lipid membrane. In other words, the set of chemicals became compartmentalised. The next section discusses briefly this most fundamental aspect of life: compartmentalisation. And whether in that set all molecules/polymers were synthesised by another belonging to that same set may not be that relevant, as the compartment could exchange materials, and energy of course, with the environs via its surrounding membrane. Nevertheless, it should be acknowledged that today's metabolism is basically self-sufficient, hence, to a large extent forming a dense, almost closed web of biochemical transformations, and whether there is complete closure or not it may be a matter of taste.

It is of interest to note that, upon some reflection, what this type of closure here discussed implies is a fundamental aspect for survival, be it ecosystems, societies, neurons, polymers, or life altogether: entities have more chance of survival if they are in a well-connected set where one entity (molecule, animal, or person) helps another (synthesises, nurtures, or educates), for in the same manner that an isolated chemical reaction will disappear as soon as its reactants or energy fade, isolated neurons will perish too (recall the abundant observations on the trophic effects of neural communication making neurons grow), or an isolated individual in a society/ecosystem will become extinct. Closure thus seems to be present at all levels, from the molecular to the societal. The similarity between the study of life and consciousness that has been noted in several parts of this text has encouraged the proposal of "neuroglial closure", in which another type of "closure" occurred, not of molecules in chemical reaction networks but of cell networks interacting with one another in the brain, resulting in consciousness (Perez Velazquez, 2020).

9.2 Biological Compartmentalisation: Good Borders Make Good Neighbours

Now that we have seen how biochemical networks, where basically almost all components are synthesised by other members of the network setting up a sort of connected chemical system, emerged out of the complex mess of a myriad of chemical reactions in the primaeval Earth, the next step that is very probable too to occur is that some of those networks become isolated in "packets"—well, not really insulated, because a totally closed chemical system is bound to fade unless matter and energy can be traded with the surroundings, so they have to be open to some degree. Thus we come to the next central step in biological organisation, that of compartmentalisation. Why is this a quite probable step?

Compartmentalisation in biological systems starts normally by aggregation of lipids forming micelles, then cells, and at higher levels modules of connected cells appear. Looking at contemporary biological cells, one can see they are dynamic compartment systems; not only the cells are surrounded by a membrane, but also the cell's interior contains other vesicles surrounded as well by lipid walls —a most important one is the nucleus, where your genetic materials lie. The outer cell membrane separates the internal volume from the external medium through a boundary, which controls the exchange of matter and energy between the cell's interior and the environment. It is thanks to compartmentalisation that we can think: it gives rise to cellular excitability or how neurons are able to fire action potentials to communicate among themselves as explained in Sect. 1.4, thanks to the differences—the gradients—in the concentration of ions between the exterior of the cell and the interior; without these gradients, there would not be neuronal activity, thus no thinking. Since such compartmentalisation is a fundamental principle of all forms of life, it is natural to ask what drove this phenomenon to occur.

The fact that it is a very probable phenomenon you can see for yourself conveniently at home: add a drop of cooking oil to water, shake it a bit, and witness, almost immediately, the formation of lipid droplets—vesicles of all sizes will float in the water. Turns out lipids (fats) like to bundle together when they are placed in an aqueous medium. It is a process thermodynamically driven, or in simple words, a more-than-probable process: the moment fats encounter themselves, they form aggregates in the shape of micelles, which are round structures where lipid molecules are densely packed. It is likely then that some chemical reactions floating around in that primordial soup became entrapped inside micelles (and here we are simplifying things enormously), and those trapped reactions, being able to exchange materials with the surroundings and owing to their self-sufficiency by virtue of catalytic closure (as we saw above), started to replicate. Apologies to those more informed for simplifying so much the scenario, as we could have talked about what molecules are more adequate for replication—that is, the nucleic acids (our RNA and DNA) or maybe some proteins—or how the micelle division took place in that ancient world, and so on, so we must skip a myriad of details and focus on the global scheme; which is totally acceptable, as scholars like Martin H. Fisher and William

James advised (respectively): "Knowledge is a process of piling up facts; wisdom lies in their simplification" and "The art of being wise is the art of knowing what to overlook". So here, we are simplifying a lot and overlooking many things, but those with a thirst for knowledge can find a great number of texts with details on these subjects.

Having said this, and for the sake of completion, the next paragraph enumerates, in scientific words, five well-known principles about the origin of life (Eigen & Schuster, 1982), and those readers not interested may skip it.

1. Formation of polymers, chemical reactions involving carbon and organic materials starting to create more complex molecular structures
2. Competition among chemical reaction networks, selection of polymers via self-replication
3. Evolution of the quasispecies (self-replicating molecular entities) towards optimal structures
4. Cooperation among competitors via catalytic hypercycles that generate new products
5. Compartmentalisation of molecules or organelles in separate areas in order to perform their specific functions more efficiently, which in turn allows for efficient evaluation of the relevant functional properties of those new catalytic cycles and molecular structures

So from simple micelles trapping chemical reaction networks, the succession may have continued to protocells and then to cells equipped with a cell plasma membrane that besides lipids incorporates proteins which are used for several functions. Perhaps the most important of these functions is that some proteins form pores and allow the flow of nutrients and substances in and out of the cells, or alternatively, the protein may not form a strict pore in the membrane but may translocate compounds, a process that normally requires energy. Whether we decide to call the primordial networks entrapped inside micelles or protocells biochemical or just chemical networks is a matter of taste, the boundary between the living and the nonliving, as we have seen, is extremely fuzzy indeed. In any event, life is a property of the whole system; one single (bio)chemical reaction alone may not have too many properties of life, but the whole network present inside our cells starts to have many more features of life, so just like consciousness is a global property of the whole individual equipped with a nervous system, it is the complete set of (bio) chemical transformations that is "alive".

In the final analysis, then, this is once again the previously mentioned tendency to approach the most probable state, that hard-to-reach equilibrium. This natural tendency to reach the most likely state has appeared so many times in our account of the origin of life that one could think that we are the offspring of probability, of the tendency to reach the most probable state. Well, it is a way to look at it. And perhaps it is a better way than thinking in terms of organisation or complexity. The next section will explain why these notions, so much used when addressing the evolution of life, are so ambiguous that may not be too meaningful when characterising living processes. In any case, it is thought that, as mentioned in the previous section, life started from a very "complex" chemical set of reactions that crystallised into other "simpler" giving rise to the early metabolism, here complexity and simplicity denoting many and fewer, respectively, chemical reactions. And one can invoke organisation too in the sense the "crystallised" chemical networks show the aforementioned closure, so perhaps it is a matter of organised complexity. But again, we shall see soon why these concepts of complexity and organisation may be very misleading when applied to the evolution of life. Just one intriguing note: it is said that the brain starts complex too, with a very large number of neurons and contacts (the synapses) that are pruned during the infant's development (yes, you have more neurons as a baby than as an adult, about twice as many, yet that does not help better thinking. Re-read Part I if you have forgotten why). Therefore, some readers may have already noticed a somewhat fallacious thought, that complexity entails large numbers—the larger the number, the more complex a system is. Accordingly, then, the early metabolism and early brains are more "complex" by virtue of larger numbers of constituents, yet in both cases these components are trimmed, so to speak, giving then rise to real life and intellect. This fallacy and the related about entropy applied to life are addressed in the next section.

To end this section, let us mention that all these processes we have talked about, the most important building blocks of life—metabolism, genetic materials, cell envelopes, and compartmentalisation—may have come into being not successively, but more or less at the same time. Here, we may have placed a sort of time frame starting with chemical reactions that became a metabolism occurring before compartmentalisation and so on, only to be clearer in the explanations, although most likely all these events happened at once—those five points a few paragraphs above.

Can we ask then when life emerged from mere chemistry? We can, but harking back to Sect. 7.2 where we considered its equivalent "when does consciousness emerge from mere perception" and we saw that this is a dichotomy that loses significance after understanding that perception is already a feature of consciousness, hence there being no real emergence from perception, then it is the same in the case of life: there was not an abrupt step when a set of reacting molecules became "alive", because chemistry is life. It is true though that there are features of these metabolic pathways and living processes in general that are not present in "mere chemistry" that you may have in a laboratory test tube. But the size of the test tube and the number of chemical reactions, in the case of Earth, were huge and very numerous, respectively, which, added to the extremely long time available for these phenomena to occur, makes the proposal for any special, abrupt step in the transition from the inorganic to the organic a difficult one indeed.

9.3 Clarifying the Entropic Fallacy

It is important to consider the topic of this section because these fallacious thoughts are the source of a misplaced sense of mystery in the development of life and the evolution of organisms. Foremost, entropy and its associated famous laws have been the source of much misunderstanding. Most likely many readers will have come across the term entropy, will have heard it denotes the disorder of the system, will have learnt it tends to increase as systems become more disordered, and will have become acquainted with the notion that life is a struggle against the second law of thermodynamics, that living processes violate this second law. This text, not being a thermodynamic textbook, will not cover the many technicalities about what entropy really represents and the use and abuse of this concept in all possible fields, from chemistry to biology and social dynamics. Let us just start asserting that entropy is not really a measure of disorder or chaos as commonly taught—although in some specific contexts entropy is reflecting disorder, this is not always true—, rather it represents energy distribution, so it is convenient to think about it as an index not of disorder but of energy distribution or dispersal (Lambert, 2002). And this is the only thing we will mention about entropy, because it is, since its creation in the nineteenth century (was coined around 1865 by the physicist Rudolf Clausius), a most debated and complex notion, not even scientists agreeing on what it is and is not—sometimes being the source of heated discussions, and sorry for this pun (entropy was originally devised as a measure of energy dispersed becoming heat, so perhaps those entropy hot disputes are changing the entropy of the scientists themselves!).

A second point that needs clarification: the second law of thermodynamics, which in its most popular form dictates that entropy always increases—and if one thinks of entropy in terms of disorder/disorganisation, then life processes being more organised imply a decrease in entropy—, does not apply to open systems far from equilibrium; it is valid only for closed systems (isolated ones that do not exchange energy/matter with the surroundings). We saw in the preceding sections that living processes take place in open systems and are not at equilibrium; hence, life does not go against the second law just because it cannot be applied to this case.

Your body may well be decreasing your entropy, but if you bother calculating the entropy change of your body plus your environment with which you exchange matter and energy, you will see entropy increasing. So, please, never, never apply the second principle to an open living (or nonliving) system without considering the surroundings. Therefore, living processes are not violating any entropic principle, they are not feeding on negative entropy, as the famous book *What is life* written by Nobel prize Erwin Schrödinger claims, something that has caused wide misunderstandings and confusions among scientists and laypeople alike, because it does not matter how many Nobel prizes one can have, sometimes your thoughts are just plain wrong.

Similarly, it has been maintained that evolution, usually described as a process which involves transition from disorder to order, or to more organisation or complexity, goes against that famous second law because order increases, hence entropy decreases, but we have just said that entropy is not a measure of disorder and that to apply the second law one must consider the system plus environment. Hence, evolution does not violate any law either.

The best manner to think about these things is in terms of probabilities instead of notions like entropy or complexity: it is all in the tendency to approach the most probable state, the probability of a final state is higher than the probability of an initial state. Although it may seem like that the evolution of higher life forms represents lower probability than just having a messy non-living system around, remember that most of Part II has been focused on the understanding that the ingredients in that recipe for interesting and "seemingly improbable" things to occur guarantee that what seems unlikely is many times very probable. All patterns we see in nature are very probable; otherwise, we would seldom see them, and life is a pattern of organised activity.

9.3.1 And Furthermore, Clarifying Other Closest Relatives of Entropy

Just above, the words order and disorder appeared—very close relatives of entropy—referring to increasing order or organisation during the course of life and evolution, increasing complexity. But is it true? A moment of reflection is always advised when using terms in a pure intuitive fashion, because we all know intuitively what order and disorder are. But what really is order/disorder? A usually offered definition of disorder is "a state of confusion"; and what is confusion? Doesn't it depend on your knowledge, the more you know, the less confused? It may be clear, for instance, what an ordered, or organised, sequence of items in a row means; this is a sort of quantitative order. If one piece of the chain is missing, we can claim it is less ordered than when all items of the chain are present. But speaking qualitatively, as in what an ordered or organised organism is, things become vague. Are we encountering here other relative and ambiguous terms, like those discussed above of probability and randomness? If the concepts are unclear, discussing their application to life will never end.

Concepts like complexity, order, and organisation constantly show up in discussing life. We'd better not touch the concept of complexity, because as occurs with entropy, there are so many notions and so many ways to compute complexity that it is altogether confusing (even for scientists) and *meaningful only if applied to a specific situation*, becoming ambiguous and leading to misinterpretations of complexity when talking in broad terms. Let us clarify that the popular lay notion of complexity as describing a situation that is difficult or challenging (how many times it is said that after a heavy snowfall the situation is "complex") has little to do with the scientific connotation, or better yet, the plural should be used: connotations. Readers not versed in these themes will be dazzled to know in how many totally different manners entropy or complexity of a phenomenon can be computed. Entropy, like complexity, is a sort of arbitrary measurement that can be applied to many observables, and hence, the interpretation varies. This situation does not help in creating a coherent framework. It is not that those many ways to study entropy or complexity are wrong; it is that they are right only in the specific circumstances of our experiment or observation. Complexity and entropy and many other concepts have several definitions and connotations, depending on context and the observables that are examined; extrapolating or generalising to other situations and phenomena has to be done with extreme caution. As advice to lay readers who are enthusiasts of entropy and/or complexity, when reading studies using entropies/complexities, one has to understand what of the many entropies the study is using, the situation where it is applied, and finally the observables used, and only after these careful considerations can one interpret adequately the results of the study; to those who do not want to take this trouble, the counsel is to avoid reading about entropy or complexity. Wild extrapolation is what happened in the aforementioned fallacy about life going against the second law of thermodynamics. And this confusing situation happens too with this other most popular concept today in the natural sciences, another relative of entropy: self-organisation—this self being a different one from that studied in Part I!

Thus, it is said life involves self-organised processes. But when is a process organised or self-organised? In Chap. 8, we saw the separation of steel balls in the two sides of the container, and it is considered self-organised in the sense nobody used their hands to fill up one chamber and empty the other, although someone had to shake the container; or in the Bénard convection, the oil molecules started to move without anybody directly imposing that pattern, but someone had to heat the oil of course. What is the self in self-organisation? Students of self-organisation have to plunge into many fields like thermodynamics and its close relative statistical mechanics, chemistry, and of course biology where self-organisation thrives, from the bird's V-formation we met in Fig. 7.1 to biochemical networks like that of Fig. 9.1. Self-organisation is a keyword in contemporary science. Those interested can find a seminal (and quite technical) article on the topic where the authors used sand piles to characterise and explain the phenomenon of self-organisation, so you can do it on the beach as the authors themselves admit! (Bak et al., 1988). For fairness sake, it should be noted that a field of science, called synergetics, has been developed as an interdisciplinary scientific field examining the formation and self-organisation of patterns, and here is another seminal work: Hermann Haken's *Synergetics, an Introduction—Nonequilibrium Phase Transitions and Self-Organization in Physics, Chemistry, and Biology* (Springer Verlag, 1983).

In short, those processes that reach a steady state all by themselves without anyone explicitly shaping it are said to be self-organised. In that sense, life processes are considered to be self-organised. We have also seen that organisation is dependent mostly on the nature of the interactions among the constituents that form the system, and it is not so much dependent on the nature of these components, so a biochemical cascade and the Bénard convection—the former within the animate and the latter in the inanimate realm—can be equally considered self-organised.

But once again, let us reflect on what organisation (self or selfless) denotes. Like before in the case of order, intuitively, we all know what this means; however, when we start to think deeply about how we determine something is organised or disorganised, we see that, just like beauty is in the eye of the beholder, organisation too. W. Ross Ashby, psychiatrist and cybernetics pioneer, reminded us that "Organisation is partly in the eye of the beholder [...] There is no property of an organisation that

Fig. 9.2 Animal mimetism. Spot the moth among dry leaves, an insect that has developed an optimal corporeal organisation for its context. (Courtesy of Project Noah, www.projectnoah.org/missions/8015991). Organisation is a relative term related to the specific conditions and context

is good in any absolute sense; all are relative to some given environment" (Ashby, 1962).

Precisely, organisation depends on the environment, the situation, the surroundings. It is all the same with all the concepts we have seen in this section, entropy, complexity, order and disorder, and organisation, which all depend on the context. A clear illustration of this point is animal mimetism; when we observe the many times amazing ways animals perform a mimicry of their environment in order to enhance their survival rates by looking similar to some environmental elements, there may be a feeling of awe, for how can this animal have perfected to such degree its resemblance with its surroundings (in Fig. 9.2, if you scrutinise the picture, you will see a moth among dry leaves). These are examples of well-organised animals for their environment, but now you place them somewhere else in a different environment and that organisation is useless to them: they will end up being eaten by a larger beast.

These wonders of animal mimicry represent an illustration of how a certain bodily organisation is appropriate, organisation driven by the genes that mutation after mutation (and these mutations can be thought of as the fluctuations in the genes, remember variability is omnipresent in all natural phenomena), after many trials and errors, successes and failures—but we almost always observe the success stories—find that some organisation fits the purpose of enhancing the survival of the creature. And one wonders how many mistakes are behind that successful organisation and how many mutations resulted in extinct individuals. By the way, mimicry does not end with looks, with the body; some animals like burrowing owls mimic the rattling and hissing sounds of rattlesnakes to frighten off predators, so perchance we can call this one dynamic organisation as opposed to the structural organisation of mimetism. And whether we call it self-organisation or just plain organisation does not change the fact that, in the final analysis, it all depends on context. Thus, a general take-home message arises: always think globally, considering the whole situation and environs, before you emit a conclusion or judgement on some specific phenomenon or event; this applies equally if you calculate entropies or if you judge your neighbour's behaviour.

In closing these sections on the entropic—and its relatives—fallacy, perhaps a bit too technical for some readers, we have seen that living processes neither violate any entropic principle (the second principle more specifically) nor are more organised than required by the surroundings. We have also seen that the essence of the emergence and evolution of living organisms is the tendency of natural phenomena to approach the most probable state, known as equilibrium. Being the offspring of probabilities rather than of a creator is not that bad. On the contrary, we can consider ourselves lucky to have won a prize in the lottery of the living. This consideration about whether our progenitor is chance or something else more involved in our affairs leads us to, very briefly, deal with yet another fallacy.

9.4 And Yet Another Fallacy: (Wo)Man and Machine

In addition to the several fallacies which this book has tried to rectify, there is another common fallacy that muddies the present theme of comprehending life and why we ended up on the surface of this planet, so let us not stop and let us deal with this misconception too. This one is to consider human bodies, and those of other animals for that matter, machines.

This leads to misunderstandings because if our bodies are machines that have been so perfectly constructed, then someone or something must have built us. In mid-twentieth century Spain, young children learned this ditty "No hay reloj sin relojero ni mundo sin creador" (there is no watch without a watchmaker, there is no world without a creator). Whereas machines are created by design, prescribing the composition and relations of the parts, we have seen in the preceding sections how patterns in nature and thus organisms can emerge spontaneously through dissipative processes without prescribed instructions. We saw that living systems are dissipative structures because they are maintained by flows of energy, and by virtue of the recipe presented in Chap. 8 which includes as well variability and situations far from equilibrium, improbable, "perfect" patterns may emerge spontaneously, the structures arising from within the system, like the Rayleigh-Bénard convection of Fig. 7.2. The structure of our (body) parts and their relations are influenced by the conditions, the contexts in which the organism develops, but in the case of machines, the structure is designed originating from an external source—usually an engineer.

That we are not really machines one can see by inspection of our bodies and those of machines; machines can be easily taken apart, but you will not be able to separate your parts, say the vascular system and the nervous system, because in organisms, the boundaries between parts are very fuzzy and taking one apart normally means destroying the system. As well, the function of machines are clear-cut and established by external design, whereas the function of dissipative structures like our body is intrinsic, emerging from their time evolution without much planning. Our bodies did not evolve by serial aggregation of parts like a machine is built. In the course of evolution, there were spatio-temporal correlations that determined whether a body part appeared or became obsolete (like wings in penguins, which became flippers). And always remember that there was a long, long time for those correlations to be evaluated. Much of the story of the evolution of organisms is explained in many texts, one that is not too technical is Richard Dawkins' *The Blind Watchmaker*, an appropriate title because evolution is blind, and any purpose we see in it is given by our psychological need aforementioned, the necessity to find an agency (an agent that makes things happen, recall that the notion of agency was explained in Chap. 2), to find reasons for things around us, for us to exist, for life in general.

To conclude, we seem to be the result of chance, and necessity too as the biochemist and Nobel prize winner Jacques Monod already told us in his *Chance and Necessity* (*Le Hasard et la Nécessité*, Éditions du Seuil, Paris), a text on how life can be explained without having to invoke final causality.

9.5 The Dance of the Genes

Having established we are not machines, can this help somehow our inquiry about the nature of us being here? Well, in spite of the previous words, we and the rest of organisms, just like machines, have a sort of "creator", perhaps not a very good one on account of the many errors that have been made through evolution. The creators are the genes. We are the product of the genes, in combination with the boundary conditions imposed by the environment. Already in Chap. 6, we met Robert M. Sapolsky saying that "Genes don't make sense outside the context of environment". Unlike machines, as stated in the preceding paragraphs, our bodies did not evolve by serial aggregation of parts like machines are built; we were made intrinsically, so to speak, from within, correlating the doings of the genes with the situations out there in the environs. It is the surroundings that judge, as it were, how genes perform their function, establishing an entangled loop between environments and physiology.

We thus go back to the topic of variability and fluctuations. These are other important fluctuations—most fundamental for the rise of the living organisms—at the molecular level that took place. Genetic fluctuations—they are called mutations— that manifested themselves as corporal conditions which were then "evaluated" by nature. We now arrive at the subject of natural selection, the evolution of organisms. And as in other topics discussed above, here we will only touch on the most essential aspects and will skip details. The evidence of natural selection can be seen all around: organisms tend to be well adapted to their environments. The essence is simple and not unlike that underlying the previously considered phenomena of the emergence of a connected metabolism and its encapsulation into lipid membranes, so we will use Manfred Eigen's words when talking about chemical equilibrium in his voluminous book *From Strange Simplicity to Complex Familiarity*: "the result of a microscopic lottery viewed at a macroscopic level". Remember the simile used in the introduction to Part II that our presence here and now in our current shape is the result of one lottery number, because one number must always win in a lottery draw. Genetic mutations occur at random, and some will benefit the organism, others will be deleterious, and yet others will pass without pain or glory.

The reasons why genetic materials mutate are many, from the solar ultraviolet radiation to mistakes made by the molecular machinery for gene replication; and because these occur randomly, thus, a sort of microscopic lottery is being played all the time and the final result we see is the macroscopic living organism. Once again, let us be fair and precise; there are some who think not all mutations are random, and there is a field called epigenetics which studies how the environment changes the works of the genes, but without necessarily causing mutations. And continuing being precise, Eigen's quote above was about the nature of chemical equilibrium, so we could add his description to the comments on the notion of equilibrium of Sect. 8.1.1, and the idea is similar now applied to the case of mutating genes and their effects on the survival of organisms: from molecular, microscopic pandemonium, we reach a macroscopic order, in this case it is the development of organisms establishing an ecosystem.

A common and very ancient misconception was (or maybe still is among some people) that new genetic variations can be forced to appear if the organisms are subjected to certain environments time after time. In other words, the environment gives rise to, directly imposes, changes in animals. This is the famous theory attributed to the zoologist Jean-Baptiste Lamarck, although these ideas precede him so that he is not really the original advocate for the theory of inheritance of acquired characteristics. In simple terms, it is not the case that if you place crocodiles to live in Antarctica for many generations this will force them to develop a cover of hair like that of a bison, but what may occur is that a mutation suddenly appears that causes the reptiles to start growing hair, and thus, these individuals will be better protected against the cold and may end up dominating the ecosystem, and yes, you may find, after several aeons (won't happen in a couple of generations!), hairy crocodiles.

Whereas Darwinian evolution is normally associated with the beneficial mutations favouring the survival of the fittest, it should be acknowledged there is another theory that proposes that some variation (mutations) may not affect the fitness of the organism and yet be selected. This is the proposal (in simple terms) of the neutral theory of molecular evolution, which advances the idea that most of the genetic variation in populations is the result of genetic drift and not of selection. It may be possible, why not, that if a gene is not causing harm to the organism, it may stay there and be passed on to the offspring. In any event, the neutral theory is not an alternative to the theory of evolution, rather an elaboration of a part of this theory—and the evidence discovered through investigations of the neutral theory also supports the theory of evolution. Always remember that the genes are only "interested" in becoming immortal, in passing from one generation to the next. And all this is done, as Dawkins pointed out in his aforementioned book, blindly. It is understandable that people through the ages have been enthralled by the macroscopic results of this molecular lottery (continuing with the previous simile), the emergence of organisms so well adapted to their environments. For instance, have you noticed that herbivores have sideway-facing eyes and carnivores forward-facing? Well, not quite, it is the prey animals which have eyes on the side, not all herbivores have this feature because not all are preys, and some carnivores do have it (like birds which are also preys themselves) because this affords a wider field of view good to spot predators in front of them or coming from the sides; naturally, this does not apply to animals that are herbivorous but normally are not prey, like sloths or gorillas, so genes giving them eyes situated like those in ducks will not be especially relevant. Indeed, the genetic lottery is the source of amazing prizes, the currency being survival.

Some genes, therefore, seem to achieve a good deal of success in becoming almost immortal, and they sometimes seem to follow the wonderful rule "if something works, keep doing it" (we humans seem to be forgetting this precept, always thinking new things to "improve" whatever is already working, many times with irrelevant or even worse results, let alone the time wasted), because once a design of a certain feature is found by the evolutionary process, this characteristic is applied to many distinct functions. But again, one should not let these anthropomorphising metaphors get in the way of the basic understanding of natural evolution, as genes have no purposes. And this concludes our extremely shortened version of the fluctuations in genes and the evolution of organisms.

Chapters 10
The Special Ones

No, we will not be talking about football, in case those fans of the "beautiful game" are thinking about a certain Portuguese coach of our times... We will talk about ourselves. Lurking in the pursuit of finding a purpose in life, a reason to live, is the belief that humans are special creatures. In what sense are we special? The sections above have already suggested that our presence in this world may be not so much a remarkable event but an inevitable one, just a matter of probabilities; as declared above in Sect. 9.2, living organisms can be thought of as the offspring of probability. We explained and thus demystified the apparent miracle of the steel balls experiment that showed how something apparently improbable is in fact very probable, we described the forces behind the beautiful Rayleigh-Bénard pattern, we examined the main causes that result in structures as varied as lightning and neurons, and concluded that the principles which gave rise to those results also give rise to all-natural phenomena around, with us included.

It is said all men are created equal, but this thought should perhaps be extended to the rest of living forms because, in addition to rising from similar Physicochemical principles as we saw, all things alive are equal in one sense: the basic essential mechanism that keeps all alive is a common one, and as strange as it may sound, the essence of life as we know it is the transport of electrons "hopping" from protein to protein on a chain. This chain of polymers transferring electrons is called the respiratory chain, also termed, for obvious reasons, the electron transport chain. Note that this respiration is not the standard one of breathing, of inhaling oxygen and exhaling carbon dioxide. This is a biochemical event that refers to the process by which we, and the rest of living forms, use energy to survive. Those ancient, primordial chemical reactions we talked about in preceding paragraphs became more specialised for the needs of living creatures, such that today basically all animate entities —human, algae or bacteria— depend on one thing, namely, the transport of electrons from one molecule to another. This is the root of all of us living critters on Earth.

J. L. Perez Velazquez, V. Nenadovic, *Being and becoming*,
https://doi.org/10.1007/978-3-030-78264-1_10

Hence, those who think we are very different from other life forms should consider a few facts and not be so much concerned with looks. Because we certainly look different from a paramecium, but our biochemistry is very similar, at least in the basic aspects of energy generation and storage or nucleic acid mechanisms of replication. And our companions the greens —the plants, those creatures about which we think little and mostly for nourishment purposes but that have extraordinary features such as communication channels among themselves— also share biophysical/biochemical phenomena with us. To wit, they too have the respiratory chain and as well they are, if one may say, a bit superior to us animals because in addition they have photosynthesis, which is another electron flow through a chain of molecules that allow them to generate their sustenance. To be a bit more precise, while photosynthesis requires energy and produces food, cellular respiration breaks down food and releases energy. And all these processes are achieved by those particles called electrons moving from donor to acceptor in the electron transport chain which is nothing more than a series of molecular complexes that transfer electrons from electron donors to electron acceptors (we shall skip the molecular details of these donors, any biochemistry textbook will contain these minutiae).

More technically, in animals this is called oxidative phosphorylation, the last step of cellular respiration, and in photosynthetic plants the technical term is photophosphorylation. Photosynthesis and respiration both consume and create the same substances (water, glucose, oxygen, and carbon dioxide) but in different ways. In photosynthesis, the energy of sunlight is used to free electrons from the chlorophyll pigments that collect the light and at the end of the chain, oxygen is created as a by-product. We, on the other hand, have to use that oxygen in our cellular respiration to generate energy in the form of a compound called ATP (adenosine triphosphate), the main energy currency of cells, and we get our electrons mostly from the sugar we eat, oxygen being our electron dump. So in the final analysis, the challenge for all life forms on Earth is the same, the biochemical/biophysical meaning of life, so to speak, is that one has to find a source of electrons and a place to dump them to complete the circuit. Isn't that something?

In short, after all we have seen in this book so far, it can be appreciated that biochemically we share our essence with all organisms on this planet, that biophysically we are subject to similar phenomena as other organisms both living and even nonliving, that psychologically we share many aspects with other animals... But neither are we completely equal with other life forms nor are we all created equal. Even identical twins are not totally genetically identical. Yes, those fluctuations in genetic materials again (these twins arising from the same fertilised ovum should have identical genes, but in reality there are some differences). Again, the term variability should be highlighted. If equal is seen to be the same as identical, then equality is bad for survival. For instance, at the time of the writing of this book, the coronavirus pandemic (known as COVID-19) is happening, and it is the unequal response of humans to the virus that will ensure the survival of the species, with this or any other infectious disease. There is wisdom in the old adage "variety is the spice of life".

The point is that our differences should be understood and acknowledged, recognised for what these are, and respect all these many disparities because these variations make humanity thrive. However, in the social contexts in which we live, the difference is not always celebrated. When there is competition for resources the concepts of "us" and "them" are amplified. This leads to discrimination and racism. Knowing now that our neural circuits are thus predisposed, education is the weapon with which to fight this ugly side of humanity. Racism, or violence of any type, will not be eradicated by demonstrating against it or by using politically or socially correct euphemisms. All these semantic quibbles and sociopolitical finesse will not help the issue. Racism will disappear once all people receive a good education and are informed about what makes us different and similar, and accept these differences without further over-elaborations. In Nelson Mandela's words, "Education is the most powerful weapon which you can use to change the world", echoed by Maria Skłodowska-Curie's "You cannot hope to build a better world without improving the individuals". Education in terms of understanding (not only memorising as is usual in schools) our past history, accepting that the past cannot be judged with today's views, comprehending the reasons for past events —because as the adage goes, those who forget history are condemned to repeat it— is the only hope for a fully rational and fair humanity.

Let us put an end to this digression and get back to our theme. What is special about us, other than our belief that we are? We do have one aspect that is really special. Unamuno's quote at the beginning of the book already gives us a hint, in that we can not only cast doubt on the validity of our reason but also on the validity of the tendencies the genes carved in our brains. Homo Sapiens has been endowed, courtesy of our genes, with a big brain, equipped with that feature called intelligence (about which we devoted some words in Part I) that may go against the will of the genes. Hence, what is special about humans is that the possibility has appeared for the first time in nature, thanks to the rise of highly evolved intellects, to go against the forces of the genes, to perhaps enter a new era where it is not the transmission of genetic material but the transmission of ideas, concepts —memes is the general term. Memes are units of cultural information that spread by imitation, so these can be ideas, behaviours or styles that become fashionable and often carry symbolic meaning. And because these self-replicate (from individual to individual), and can mutate too according to selective pressures (for instance, pure and hard slavery of the old times giving rise to modern slavery at work), therefore memes can be the cultural equivalent to genes —apparently the term was coined by Richard Dawkins in his book *The Selfish Gene*. In point of fact, we have cultures because people stick together sharing similar conventions, a manifestation of the power of the transmission of ideas, the influence and ruling of the memes. Now there is some food for thought here, because you may have little choice to inherit, say, a bad gene that will predispose you for an illness, but you may have more choice to avoid becoming infected with a "bad" meme. Or perhaps not? Oh well, this is again the topic of free will that was so much discussed in Chap. 2 and others in Part I.

Nonetheless, our powerful brains can go against the will of the genes. As such, whereas genes, only caring about passing from organism to organism, would dispose of unhealthy individuals, we, on the other hand, thanks to using our cognition, take care of the ill and disabled. As well, genes would not applaud homosexual relations, and yet we accept these, at least in many parts of the world. Hence, we have already started to alter the tide. The crucial point is whether we will be able to curb the innate forces that the genes have sculpted in our brain, neural circuits that make us be aggressive, wage wars, and have related bellicose behaviours. Charles Darwin himself proclaimed, in 1858, that "All nature is at war, one organism with another".

It has been emphasised in this book that our behaviours are on a continuum with those of other species, we just rationalise our genetically-driven tendencies with ideas; ideology makes us tolerate and excuse some of our worse habits and behaviours. There should not be any problem in understanding and accepting that our behaviours have been planted, so to speak, by the forces of nature; that the genes have moulded our brains and rest of the body to survive better than our neighbour; that those inherent forces compel us, like the rest of living forms, to perform actions that while meaningful in the wild, may not be that meaningful in our created societies. Humans have the potential to change (to some degree) the course of evolution. Remember that in Part I we examined the neuroscientific essentials of our actions, how novel external inputs enter the brain circuitries adding themselves to those never-ending chains of neural activity, carrying with them the potentiality for transformation, for behaving in a not-so-genetically driven manner. Can this potentiality be taught, improved, developed?

To some extent, yes. There are methods that foster what can be called mental education, and bring about the possibility for the changing of behavioural dispositions (more on this in Part III). Even politicians know this! In the preamble of the UNESCO Constitution of 1945, it says "That since wars begin in the minds of men, it is in the minds of men that defences of peace must be constructed". Well, politicians may be aware of this, but how much they are putting into education is another matter, perhaps due to their short-term views that normally do not go over the next election time: in the US there is way more expenditure in the armed forces than in education, but at least in the EU the trend is opposite. The philosopher Hegel pointed out that "What experience and history teach is this – that people and governments never learned anything from history, or acted on principles deduced from it". Now, almost two centuries later, his appraisal of the short politicians' and general public memory span continues to be accurate.

As highlighted a few paragraphs above, good education is a fundamental activity that will bring humanity to the next stage. It will not be technology, although we are welded to technology and are awed by its rapid advancement. It is not the quick progression of space exploration, the remarkable advance in microelectronics, the genetic engineering miracles, all these technological wonders that make us think we are on the edge of the next stage in our evolution, by themselves and without proper education they will only create modern, technologically advanced troglodytes. But to offer proper education is not easy at all. In our societies, we consider intelligent those who can solve mathematical equations or individuals who perform science to a high degree, and yet for me, it is the teachers, the educators of our youths, those who have to face crowds of youngsters every day and are able to teach them, to make them understand today's lesson (and not only memorise it until the next exam and then forget), who are the real geniuses, and not the mathematical wizards. Future will tell what we do, even though in the end our achievements and what becomes of us may be another inevitable result.

In any event, while it is true that man is not the paradigm of mercy, the tendency is clear: the world is becoming a safer place. Perhaps you disagree if you watch, read or listen to the news, but make no mistake, it is a much, much safer place. It has been documented. Crimes have been quantified because police records do exist dating back centuries. There are fewer crimes now. Only a little more than one hundred years ago, while travelling between the cities of Oviedo and León in Northern Spain, the chances of being robbed and assaulted crossing the fields and mountains between the two towns were very high. And those who like reading history will know that characters like Julius Caesar or Napoleon could barely go 6 months of their lives without fighting a battle. It is true that the twentieth century saw tremendous wars, it is true that there are still massacres and brutalities, but today's planet is much pacified, with some small wars here and there. Although the fact there are fewer wars now may not be so much due to the power of our intellect as to the influence of fear —courtesy of our nuclear arsenals— because why wage a big war where victory is equivalent to defeat. Perhaps we have become better because life in our societies requires individuals to overcome ancestral genetic inheritance that drives us to aggression, and not so much because our generosity and mutual understanding have been enhanced by the use of our logical powers.

One way or another, the world is safer at the time of this writing. We will not try to convince you, rather advise you that in case you still doubt it, in addition to stop watching daily news, to read Steven Pinker's *The Better Angels of Our Nature – Why Violence Has Declined*, where you will find data demonstrating declining violence and possible reasons as to why this is happening.

It is noteworthy to mention that whereas the behaviours needed for survival, or in other words the actions that genes make animals perform, include notably selfishness and aggression, it is us, Homo Sapiens, that have gone one step further and from behaviours within a context that have adaptive value (for instance, if you are a lion male it is natural to be selfish enough to be the first of your pack to have a morsel of the prey) we have created actions out of context, to wit, violence. Because predation just like animals do is natural in the sense it has adaptive value, but we have transformed that into violence, which is escalated aggressive behaviour expressed out of context. In the classic paper *The logic of animal conflict* written in 1973, the famous biologist J. Maynard Smith and his co-author G. R. Price already teach us that conflicts among animals, at least those belonging to the same species, are of a "limited war type" (the authors' own words) in that most fights are rituals involving tactics that seldom result in serious injuries, let alone the death of the opponent (Smith & Price, 1973). Those who watch nature shows will probably have seen goats or deer fighting, where the fight is a matter of pushing and crashing their antlers. But homicides, genocides, and other traits with which we humans entertain ourselves from time to time are, if not completely absent, at least very uncommon among animals. So the criminal invertebrate of Fig. 10.1 is probably a very large human disguised (although, to be fair, some animals do have some practices like ours, for instance, some ants —blood-red ants particularly— practise slavery, going on invading trips to find slaves among other ants). Killing for the sake of killing is extremely uncommon in the animal kingdom. Animals have a "natural" reason to kill another animal, humans have other types of, quite unnatural, reasons. It is not without interest to see that the same brains that can make us improve our lives and societies by opposing to some extent the forces of the genes, can also transform us into the cruellest villains on the planet.

In some sections of Part I, especially Sect. 1.8, it was explained that the behaviours of us and other organisms, the responses to stimuli, lie in a continuum; from the pure taxes or reflexes to the fully conscious and voluntary actions, the purpose is to satisfy the needs for survival. But perhaps due to our intellect, that at the same time that is curbing the aggressive and selfish tendencies that nature (genes) has

Fig. 10.1 The impossible criminal centipede. It may have the genes for aggressive behaviours but lacks the intellect to become a serial offender. Natural inherited tendencies plus a powerful brain sometimes bring about unnatural actions, as explained in the text, and that is why only humans display criminal behaviours. (Drawing by the author)

given us is expanding the natural predatory behaviours into out-of-context fields, we have become something else. This is another "becoming", one that lies in that behavioural continuum. Notoriously, we have an evolved predisposition to kill members of other groups. In fact, take the case of serial killers, so much popularised in movies, it is nothing more than an extreme antisocial variant of the natural phylogenetic predation, as L. Miller said in his article "The predator's brain: neuropsychodynamics of serial killing": "it is pathological only in terms of degree, not the nature of the act, and the brain mechanisms involved are on the same continuum as those related to more normal forms of hunting, group combat, romantic pursuit…" (Miller, 2000).

And yet, some argue that extreme violence and war may not be in our nature after all. See for instance "Why We Fight", published in Scientific American, where it is considered the possibility that wars are social events that emerged from the evolved propensity to eliminate potential competitors (Ferguson, 2018) —other primates like chimpanzees make war too, but of course of very limited scale. The controversy over the historical roots of warfare is presented in this paper. In any case, whether a product of our brains or not, extreme violence arose in our societies, and now it is up to those same brains to find a way to stop it, if only because we are so detached from nature now that these natural tendencies of our genetic inheritance make little sense in our artificial ecosystems we call cultures or societies.

Consequently, it seems we are actually using our "speciality", we are going against the blind forces of nature… But, are we really? Some more food for thought: if the natural forces —that is, the genes and related aspects— have made our brains and endowed them with the cognitive powers we possess that may alter the will of the genes, isn't this tweaking of natural selection and evolution, as it were, something that was about to occur sooner or later, and thus an inevitable event that in the final analysis has been caused by those blind forces? What lies in our future? Perhaps a new type of natural selection, designed by us who at the same time were designed by the genes. Who is then the ultimate designer? If some readers find a parallel between these deliberations and those of Part I about who is in charge in the brain to determine the behaviour—the last ventriloquist—they are right. As expressed many times in this book, natural phenomena—and minds and life are part of those—are so intertwined that attempts at finding unambiguous answers to these questions by separating what cannot be untangled lead to conundrums of a large magnitude; in the words of Francisco Varela and colleagues, "Organism and environment enfold into each other and unfold from one another in the fundamental circularity that is life itself" (Varela et al., 1991).

In any case, insisting again on circumventing dichotomies, we have to be content with the thought that we are the product of the genes and the rest of natural forces, and that nothing can separate us from these, for in that case, we would cease to exist.

We would cease to exist, but not life. Because in spite of Homo sapiens' career of nature serial killer, despite our "attempts" at changing the climate and exterminating living creatures of all shapes and forms, life is, much like the sense of personal identity we saw before, virtually immortal.

Chapter 11
The Enduring Life

In Part I (Chap. 5), we examined the possibility of eradicating the sense of personal identity, the self, and concluded that this is a most difficult task. Is life as well impossible to extinguish? In this time and age with climate change, pollution beyond limits, and other man-made efforts to upset the biosphere, should we be concerned that we are wiping out life on this planet? Some particular organisms may become extinct, but life as a whole is impossible to terminate. For starters, remember we defined life by the enumeration of features, and two most general features are replication and the exchange of energy/matter, which are characteristics of some chemical reaction networks as explained in Chap. 9. Even if oxygen were to disappear from the atmosphere, there are organisms which live without it; it is known that certain microbes are able to live in anoxic environments (anoxic meaning without oxygen; hydrothermal vents are examples of anoxic environments), and this should not be surprising because in early Earth there was no oxygen, and yet, many of these lifeforms like bacteria and archaea (microorganisms similar but different from bacteria) lived at that time and are still living happily without oxygen today. In case someone is wondering how these creatures dump their electrons if they do not use oxygen, which is animals' electron deposit—as explained in the comments of the preceding section on the respiratory chain, a living organism needs to deposit electrons somewhere at the end of the electron transport chain to create energy—, rest easy because they do not have any problem; these animals can use metals such as iron to drop the electrons. Life always finds a way out; it is what we have claimed throughout these sections: massive numbers of chemical reactions guaranteeing high probability for interesting things to happen. Not only microbes but also some animals have been found that apparently live in anoxic conditions: in the deep abysses of the sea, there is oxygen-free water where unicellular animals were found, and if you care to know, they go by the name of loriciferans. Life will for sure disappear on this planet when our Sun becomes a nova and boils the inner planets, Earth included.

J. L. Perez Velazquez, V. Nenadovic, *Being and becoming*,
https://doi.org/10.1007/978-3-030-78264-1_11

The similarity in the endurance of the sense of personal identity and of life can be added to the similarities mentioned in Sect. 9.1 about the commonalities in the study of life and consciousness. Mind you, whereas the self will vanish upon the death of the individual, life will not, because he/she will become something else, other organisms will rise from the decaying corpse—funny thing, skeletons are considered the ultimate death symbol, and yet so much life has gone into making that skeleton: it is called decomposition, and we will skip details, but let us just say it is the ultimate recycling we will all diligently accomplish. Nothing in nature goes to waste (entropy allowing, and this is a comment for the professional physicist!).

The finding that organisms can live in anoxic conditions, some even in notoriously poisonous environments—at least poisonous to the rest of living things—should not be a matter of wonder if the essence of the phenomenon termed life was understood. In the words of N. Goldenfeld and C. Woese: "With a proper understanding of the phenomenon of life as a dynamical process, for example, one would not find it surprising to learn of life in so-called extreme environments (such as deep beneath the ocean floor) or even on other planets" (Goldenfeld & Woese, 2011). Part II has surveyed the current knowledge that indicates the high probability of what we call life to emerge in a place like Earth. It is, as the sentence above claims, a dynamic process that finds its roots in the massive numbers of chemical reactions derived from the special chemistry of carbon (explained in Sect. 9.1). We have seen how, in an apparent paradox, the Earth chemistry has gone from high to low complexity, in which special chemical reaction networks were crystallised and trapped in lipid vesicles that later became cells, and we saw how this less complexity, or more simplicity, is better organised than the previous immense number of chemical reactions in the primordial soup of ancient Earth. This organization favours the emergence of properties associated with life and organisms to thrive while adapting to their environments, courtesy of fluctuations in genetic materials and in general in the molecular events taking place. All this guarantees that even if life as we know it were to totally evanesce (again, impossible as it is, but let us assume it may happen just for the sake of the argument), it will re-appear again.

And for what purpose? Same purpose as originally millions of years ago. Not much more than for molecules to react, energy to spread, and genes to proliferate. There does not seem to be much more sense in life, unless we hypothesise vague postulates like universal consciousness or the existence of spirits and an afterlife (as good empiricists, we prefer to start with proven observations). Reasons that this apparent lack of purpose in life should not be disturbing to us are explained in the postscript below.

Chapter 12
And Why There Is Something Instead of Nothing

Contents

And now let us offer an answer to that bonus question promised in the introduction, a question asked since very old times (the ancient Greek philosopher, pre-Socratic in fact, Parmenides of Elea already toyed with it). It is a question of apparent great metaphysical depth but that occurs equally to heavy thinkers and to children. Why there is something instead of nothing? The "answer", as it were, to this query is that it makes no sense; it is an ill-posed question.

One must note that "there is" entails existence, whereas "nothing" implies nonexistence. There cannot be nothing, because "nothing" is not a sort of vacuum, nothing is nothing, pure nonexistence. So "there could have been nothing" is not an option. This does not depend on the language used, it is a matter of logic; for example, in Spanish, it goes like "hay algo" (there's something) versus "hay nada" (there's nothing), and again, "hay nada" is meaningless, and if one insists on adding a negation as in "no hay nada" (literally "there isn't nothing")—which is the correct Spanish way of saying it— then this makes logical sense because the two parts denote nonexistence ("no hay" and "nada"), but in the end it says that there cannot be nothingness.

Other thinkers reason too that nothingness is a human concept that is only a construct but inappropriate for a description of a possible state, or the absence of state in this case. Jim Holt in his '*Why does the World Exist?*' (Liveright Publishing) goes along the same lines to reveal the absurdity of the question, saying, "Suppose there were nothing. Then, there would be no laws, for laws, after all, are something. If there were no laws, then everything would be permitted. If everything were permitted, then nothing would be forbidden. So if there were nothing, nothing would be forbidden. Thus, nothing is self-forbidding. Therefore, there must be something." So in the final analysis, the question has no meaning; it does not make sense to consider the possibility of there being nothing.

If asking whether there could be nothing is meaningless, perhaps it can be reframed. The question can be recast as why things/events occur in this way and not in another way, which is why in Part II there has been an exploration of causes for natural phenomena to occur, for patterns of organised activity to develop and become one or another thing.

J. L. Perez Velazquez, V. Nenadovic, *Being and becoming*,
https://doi.org/10.1007/978-3-030-78264-1_12

To put an end to these, perhaps, tiresome semantic quibbles, let us mention that asking questions is sometimes not as trivial as one may think; questions have to be logical and well-posed. Other famous riddles suffer from the same sin of being unsound. For instance, another typical example of an ill-posed question is the riddle about what happens when an indestructible fortress is hit by an all-powerful rocket that destroys every possible thing. This riddle is meaningless because there cannot exist in the same universe, by definition, an indestructible thing and something that destroys everything, so they can never meet. And on the subject of ill-posed expressions, one that is intimately related to our themes in this book is that one that people many times say, "If I were you…" or "If I had been born in other times…"; hopefully, it has become clear after all we saw in Part I that you cannot be me and that if "you" had been born in another time, it would not be you but someone else, because changing your genes and your history (your memories) means you are not the present-time yourself anymore. Asking logical and meaningful questions is many times an art that can change us and the world around us. In fact, many fundamental advances in science started as questions—and in this time and age, it is troublesome to see that scientific research is becoming not so much involved with good questions about natural phenomena, but rather about the business of generating more data that will bring more funds to the laboratory to continue producing more data to obtain more moneys and so on, so the loop continues (those interested in the state of affairs in present-day academia can check it in the book *The Rise of the Scientist-Bureaucrat*).

Now we have seen "why" there is a universe, and in Part II we have understood why we are part of it. Let us finish this part with some afterthoughts.

12.1 Postscript to Part II

Notwithstanding the psychological need to find a sense of purpose in life, in this Part II we have commented on the real "purposes", those of the genes, of evolution, that are no purposes at all but rather a matter of the emergence of the most probable. We have seen how the emergence of life is a very probable, basically inevitable, event, and to understand why this is the case many words have been devoted to the concepts of emergence, pattern formation, equilibrium and non-equilibrium, ingredients of recipes for interesting things to occur. All this to demonstrate that things just happen without the need of a purposeful engineer. We have seen how from non-equilibrium and energy dissipation, adding some fluctuations in the components of the system, organised patterns of activity emerge, and among these patterns, we find life processes. We explained the idea that life evolved from very complex chemical sets of reactions that crystallised into other simpler ones, an early metabolism that originated lifeforms, and finally us. Thus, we came into being, and we become. Constantly our bodies change materials, our brains make and update models of the world to adjust to the circumstances as we saw in Part I. All this becoming granted by the redistribution of energy, the exchange of energy among the elements of the universe.

People think anthropocentrically, need to see an agency responsible for the events, need to imagine a cause of processes. It is hard to think otherwise, such as the psychological need that some people suffer, so to speak, to find a reason to live. But unless we opt to delve into mysticism, it is hard to observe any purpose or meaning in life, other than the aforesaid dissemination of the genetic materials or atoms and molecules reacting and energy spreading. The psychological necessity to find a life meaning has been recognised for a long time and is the basis of Viktor Frankl's logotherapy, a school of psychotherapy founded on the belief that human nature is motivated by the search for a life purpose; logotherapy is the pursuit of that meaning for one's life, something that can be comforting to some people, because in the end, almost everything is in our minds.

Should these thoughts about the lack of specific purpose in our lives be unsettling? Can they originate perhaps a feeling of emptiness, even futility in life? But it should not be so, for there are plenty of things around to fill up our lives. And these are not precisely the bewildering multitude of technological toys we have today at our disposal, rather the natural things, living creatures, geological formations, natural phenomena all around us including the behaviour of our fellow humans, sometimes so funny, sometimes so illogical, yet providing a sort of meaning to our lives in our atoms of community, a kind of fulfilment that accompanies our endeavours in the journey through what we call life. The contemplation and enjoyment of so many natural phenomena could keep us busy for a long, long time. Eleanor Roosevelt said once that "The purpose of life is to live it, to taste experience to the utmost, to reach out eagerly and without fear for newer and richer experience", so she already told us that the purpose of life is to live it, and that's that; to taste the experience to the fullest, to reach for a richer experience. Aristotle understood this very well when he started his *Metaphysics* with the words "All men by nature desire to know. An indication of this is the delight we take in our senses". What else needs to be there? Time, of course, a most fundamental aspect of our existences.

Time to experience things. Time, that notion we devised to measure the becoming of things, has itself become a commodity in our current world. I have always admired those people of old that took years, sometimes decades, to complete a work, to achieve a goal. It took nine years to complete the alabaster altarpiece of the basilica of the Pilar in Zaragoza. Beethoven's ninth symphony was about three decades in the making. Time is our real possession, even the poorest of men have time, and unlike the other earthly possessions, you cannot recover it. Because one may have loads of riches and moneys and may lose them, but there is the chance they may be back, at least in part, but the time lost will never be back. It is of interest how many individuals need extra help to realise this fact, for example, witnessing a fatality after an accident. Life unfolds in front of our perception at full speed, from the time when we tried to squeeze the possibilities of youth to old age when we remember the distant past that now, with the perspective given by time, seemed were such happy times. Today we are constantly with an eye on the clock. Enslaved by the schedule, we have lost the capacity to be captivated by our task at hand, and we are losing as well the capacity to laugh at ourselves.

And perhaps, this is the pure and straightforward meaning of life.

Part III
The Philosophical Perspective – How Do I Experience Reality?

It is complementary to the scientific parts of this book to consider the concepts of the self throughout history and to comment on some philosophical and religious perspectives on self and consciousness, and life in general, because as we shall see, some philosophical ideas perfectly match the scientific perspectives. So, despite Ambrose Bierce's opinion that "Philosophy: A route of many roads leading from nowhere to nothing", let us explore a bit what the humanities say about our topics. But this, not being a treatise on religious and philosophical traditions, will be a brief part, focusing on the aspects that coincide with the science we have seen before and how these views help us better comprehend reality.

Chapter 13
The Self and Consciousness Throughout History

Let us start the historical voyage through the olden ages with some curious facts. For example, do you know that the word consciousness is derived from the Latin *con (with)* and *scire (know)*? When Romans shared knowledge, they had "consciousness", and in mediaeval Latin, consciousness meant a knowing subject. And did you know that neuroscience as a word did not exist in ancient times? Hellenistic, Roman and Mediaeval "neuroscience" was *sensus communis,* a sense that unites sensations of all senses in a general sensation or perception. It can be appreciated that these early concepts are very much related to current thoughts about consciousness. Today's neuroscientists talk about higher-order association areas in the brain that we have met in Part I, like the prefrontal cortex, where this "union of sensations of all senses" may take place.

In reviewing the ideas about our main themes of some important thinkers of the past, one has to talk about Pythagoras (~550 BC), who may be considered the first dualist. Recall that at the end of Sect. 7.2, we met dualism as the doctrine that insists on separating mental from physical states. He thought that matter and soul are mystically connected and that the soul can rise to union with the divine. He placed thought processes and the soul in the brain and not in the heart as was usual in his time. Enthusiasts of universal consciousness probably should know that Anaxagoras (~450 BC) introduced the "nous" (universal mind, infinite and pure) to Greek philosophy. But not all Greeks were dualists; in fact, Aristotle (~350 BC) proposed that mind and body are intertwined, inseparable in all living beings, which is along the lines of what we have maintained in Part I. And let us not forget Pyrrho of Elis (~300 BC), founder of Greek scepticism, who already shared our comments in this book about there being no really objective knowledge, only opinions.

Around the same time in Greece, the stoics thought about self-ownership, in that all animals behave in a self-preserving manner and are aware of themselves, and aware of themselves in relation to other animals, although these Greeks did not have the mirror test we saw in Part I! By the way, today, the word "stoic" commonly refers to someone who is indifferent to pain, grief or pleasure, a notion that is a bit

J. L. Perez Velazquez, V. Nenadovic, *Being and becoming*,
https://doi.org/10.1007/978-3-030-78264-1_13

distinct from that of the old Greeks. The stoic school of thought proposed a determinism in the phenomena around and hence in behaviour too, that finds parallel to our comments in Chap. 2 on the determinism in free will.

And in the Roman world, Seneca (3 BC–65 AD), who was born in today's Spain, already pondered about possible definitions of the self while strolling in the gardens of Cordoba (then Corduba). He thus declared the self as a reference point of action without which no animal (man included) can act appropriately, like nourish oneself, recreate or protect oneself from dangers. We see that his views coincide with much of what we examined in Part I about things like agency and the self as a model created by the brain.

Hierocles (in the second century of our era) wrote *Elements of Ethics* where he mused about self-perception. Like many before him, he also attributed a sense of self to animals, arguing that birds, reptiles and mammals from the moment of birth perceive themselves continuously. Note here the similarity with our proposed concept of self as the perception of the unity and continuity in behaviour and cognition, and that self-perception is a most basic faculty of animals.

Marcus Aurelius (121–180), Roman emperor, had enough time from war to war to engage into some serious thinking. In his *Meditations*—a short compendium of his contemplations—he emphasises morality combined with emotional detachment and considers that all is being continually transformed (he knew about our "becoming"). The importance he gives to self-discipline, virtue and inner tranquillity, plus confining oneself to the present, all are reminiscent of Buddhist thought, of which most likely he knew nothing although there have been accounts of some Buddhism interactions with the Roman and Greek worlds. Two examples of his musings related to what we have talked about are "Everything is intertwined, and there is not anything that is not linked to another", and "If you are sad due to some exterior factor, it is not this one that perturbs you, but your judgement of it". And with regard to this one, "I am made from a formal cause and from matter; none of these two will disappear into nothingness in the same manner as they did not originate from nothing", he was very right in the "matter" part, but as to the "formal cause", that remains too philosophical for some tastes.

Going into medieval times, the views about life and self are mostly static. The dynamics of "becoming" are missing, but the "being" is present, as exemplified in The Great Chain of Being (in Latin *scala naturae*), a hierarchical structure of all matter and life, already advanced by Plato and other Greeks and further developed during the Middle Ages. The chain starting with God progresses downward to angels, humans, animals, plants, and minerals. Important thinkers in the medieval explorations of consciousness are Muslims like Ibn Sina (westernized as Avicenna, ca. 980–1037), who proposed the self as an immutable subject of constantly fluctuating experiences, or Ibn Rushd (Averroes, 1126–1198), who conceived of two selves, a primitive one which is simply a subject of experience and a more complex one which amounts to a rational organ. This is reminiscent of Edelman's and Damasio's views on primary and higher-order consciousness and the proto-self, explained in Sect. 1.8.

The Sephardic Jewish philosopher Moses ben-Maimon (westernized as Maimonides, 1135–1204), born in Spain and one of the most influential scholars of those times, was an advocate of the soul. He distinguished two kinds of intelligence in man: one material dependent on the body and the other immaterial, independent of the bodily organism. Due to its witty writing and its relation to our theme regarding the enlargement of the sense of self in the culture medium of our societies, we will quote him here in a script where he describes and shows concern for a most prominent human sin, that of conceit: "There is one disease which is widespread, and from which men rarely escape […] I refer to this: that every person thinks his mind [...] more clever and more learned than it is [...] I have found that this disease has attacked many an intelligent person [...] They express themselves upon the science with which they are familiar, but upon other sciences about which they know nothing [...] If met with applause [...] so does the disease itself become aggravated".

St. Thomas Aquinas (1225–1274) was another leading thinker who had at least some dynamic perspective in the themes of consciousness and life, in that he described perception as an active process rather than a mere passive receiving of sensory information. And in a sort of Buddhist-like manner, he placed no distinction between cognition and sensation: by perceiving, the perceiver becomes one with the perceived forms. This is resonant too with what we have seen in Part I about not making strict distinctions in the processes of cognition/consciousness. He also devoted some thought about free will; in his *Summa Theologica*, he writes that the intellect and the will are engaged in a dynamic and complex interaction, with multiple stages between the initial perception to the final action of the will. Nobody is perfect, as we can see here, in spite of having a non-dualistic tendency and seeing cognition all as one phenomenon as mentioned above, he seems to still distinguish and separate intellect and will —man's tendency to dissect is indeed extremely persistent, not even saints can overcome it.

Let us just mention a fellow with the bombastic name Phillippus Aureolus Theophrastus Bombastus von Hohenheim, Paracelsus for short (1493–1541), because he is credited as providing the first clinical/scientific mention of the unconscious. It is interesting that in spite of all people losing consciousness once a day (during sleep), it had to wait until Paracelsus' time for unconsciousness to be addressed scientifically.

It was generally believed in the mediaeval and Renaissance times that the mental action occurred in the cerebrospinal fluid, not in the mass of the brain. Cognitive functions were thus located in the major fluid-filled ventricles of the brain. So now we have reached the Age of Enlightenment, with prominent explorers of consciousness such as René Descartes (1596–1650), with his famous dualism, the distinction between mind and soul; for him, consciousness (res cogitans) was not physical at all. His views, as we all know, are still alive among us. And to be precise, dualism had appeared long before his time, already in ancient Greece and in Eastern traditions, but it was Descartes who gave it the definite push towards fame. It is perhaps understandable the fascination with this view of separating mind and brain, for some of us an artificial divide as has been previously explored in antecedent sections. I am not sure what mystery there is in the fact that due to the function of the nervous system in processing information, psychological aspects like sensations, feelings and other subjective experiences arise (these reflections were treated in Sect. 7.2 on the subjectivity of perception). Nobody seems to make a big deal of the results of digestion after the gastrointestinal tract processes food, but there is a lot of fuss about the results of brain processing information.

Another character whom we met in Sect. 3.1 is B. Spinoza (1632–1677), who had a perspective close to the one advocated in this book, that of embodiment: consciousness resulting from the brain inside a body immersed in an environment. In that section, we saw his views on free will. Recall that he said, "…this is that human freedom, which all boast that they possess and which consists solely in the fact that men are conscious of their own desire but are ignorant of the causes whereby that has been determined".

We could mention many more scholars, some have already appeared in this text like Hume or Locke, but since this is not a historical treatise, let us stop here and finalise mentioning views of the self and consciousness in ancient eastern traditions. There are texts of Hinduism that expound Vedic theory, which dates back to about 2000 BC. Here, awareness is viewed in terms of the reflection that the hardware of the brain provides to the underlying principle called the self. The conscious self is taken to be a reservoir of infinite potential, with capabilities determined by the hardware of the brain, this hardware representing the structure to focus the self. In short, self-awareness was considered an emergent phenomenon, very much as we discussed in Part I. These texts also prescribe the principle of complementarity, in which the material and the conscious are aspects of the same reality, again a very holistic viewpoint that coincides with much of what we know today in the neuroscience of perception and action. The Vedic model of mind is exemplified in the chariot parable (contained in the *Katha Upanishad*), a mind that is understood as emergent on the brain hardware contingent on the principle of the self.

As can be appreciated after reviewing the ideas of times past, there is not much new under the sun. These ancient thinkers, in the West or in the East, already grasped some aspects of what today is more established by virtue of our scientific technology. It can even be admitted that some of our current theories about the self and consciousness are but refinements of what the old sages thought about. Never forget that you do not need high technology to do good science... But it helps in some specific projects!

In this section on views on self and life, the perspective of aboriginals should be considered too, as they have more holistic views than the predominant Western perspectives (recall the example mentioned in Sect. 8.1.2 about giving personhood to a river). Indigenous peoples are widespread across the world and it is beyond the scope of this book to explore all. The views of First Nations of Canada, being predominantly oral cultures, may not have been taken into much consideration and yet have a distinct view of being and becoming. Of note is the view of the interconnectedness of all things in the universe. The nature of the interconnectedness of organs within the body is replicated and multiplied between people, between people and animals, and between people and inanimate things. This is a development over thousands of years of survival predicated on living within the natural environment. This survival relationship shapes the personal being and becoming as well as that of the natural world culminating in the granting of legal "personhood" to a river, giving it rights and acknowledging the river's own biodiversity. This glimpse into another journey of being and becoming reinforces that the perception of self is generated by the brain "programmed" by its internal and external environments, such that the reality that emerges, that leads to one sense of self and world view over another, is again the product of the neural circuitry. This echoes the biodiversity of the world in a plethora of selves.

Perhaps, the tradition that has the most points in common with the current neuroscientific and biophysical views we have examined in this book is Buddhism. Hence, we will devote a section to it.

Chapter 14
The Power of Contemplation: Explorations on the Self and Consciousness in the Buddhist Tradition

The philosophy of the Buddhist tradition resonates prominently with what we have seen about self-awareness in Part I and about the nature of life in Part II. It is also very much in line with our title. For starters, this is the Buddhist doctrine in one sentence: everything changes—yes, this is, in just two words, everything the Buddha taught, as the Zen monk and teacher Suzuki Roshi once declared when asked to summarise the doctrine. Being and becoming is the kernel of these teachings. As Joseph Campbell said, "We must let go of the life we have planned, so as to accept the one that is waiting for us". The plan now is not to explain Buddhism and review its practices; rather, the aim is to point out the similarities with the teachings and findings of science in the aspects that can help endure and simplify our lives so that things become more enjoyable.

The fact that contemplative methods (popularly called meditation) and science share several interests is the reason why there are institutions like the Mind and Life Institute, founded primarily by the Dalai Lama and a neuroscientist, Francisco J. Varela (1946–2001). To be more precise about these commonalities, we have explored in Part I the fundamentals of neural activity and how this leads to our brain function, our thinking and our intellect. Recall those never-ending neural chains of activations, which produce a sort of mental walk: in our everyday lives, we normally go from one thought to another, from one perception, that is elaborated and many times over-elaborated in our minds, to another. We are all aware of this mind wandering; just monitor your thoughts for some time during a normal day (of course, if you happen to be extremely preoccupied with something or momentarily busy and concentrated in an action, then this mind wandering may be diminished, but this only occurs in special moments). It turns out that the Buddhist tradition recognises and places strong emphasis on the troublesome aspects of this mental wandering, especially when we over-elaborate what we perceive. According to the Buddha's teachings, hate, envy and other undesirable traits arise from elaborated perceptions and notions. The early Buddhists may not have known about those neural loops of our brains, but they certainly were aware of this sort of human habit that sometimes hinders us from fully enjoying life.

J. L. Perez Velazquez, V. Nenadovic, *Being and becoming*,
https://doi.org/10.1007/978-3-030-78264-1_14

Thus, this philosophy highlights the importance of pure awareness of the perceptions rather than the messy interpretations and evaluations. An awareness that helps to reach equanimity and avoid agitation, which helps us live in the present rather than being concerned with the future or the past. You have probably caught yourself many times thinking or over-elaborating about past and future events and forgetting to live in the present. It can even be said that people live more in the past and in the future than in the present, that is, they live in a non-existent time since only the present exists. Recall too the comments in the postscript to Part II on our modern slavery by the schedule, on how we have lost the capacity to be captivated by tasks at hand. Whereas the nonjudgmental quality of mind which does not obsessively anticipate the future nor reflect back on the past may be helpful to reach equanimity, some thought about past and future is required to lead a "normal" life. It is all in the equilibrium, not that one talked about in Part II, this equilibrium is the commonly understood notion of neither too little nor too much, just the right amount. It is necessary for survival to follow our worldly desires but, as the family doctor usually tells us, keep the right balance, this sometimes hard-to-find equilibrium.

So, is this mind wandering a problem? It should not be a great concern in general, perchance only at times when one has to be fully concentrated on something it may pose a problem, if one cannot focus the attention properly. However, the apparently innocuous wandering of our thoughts in a reflex-like manner without much conscious control brings about quick judgments, prejudgments or prejudices, and compulsive actions. To properly judge requires serious thinking over a certain time. So many times, when our attention flickers, when our patience recedes, when reason gives way to judgmental emotions, to impulse, our actions become not the optimal ones we would have liked to perform in other, more calmed conditions.

Already in several sections of Part I, we remarked our most common routine of judging our neighbours based on a hasty encounter or look; we also described examples of the rapid activation of certain brain regions upon perceiving some people, things, or events and its consequences on almost reflex-like behaviours (e.g., it was described in Sect. 3.1 the activation of the amygdala after participants were shown faces of a different race, with the consequent possible results due to the involvement of the amygdala in fear and aggressive behaviours). It is no coincidence that the word "reflex-like" has been used in this book a good number of times. This is then the potential problem; these judgments and opinions about other people and circumstances, these impulsive actions, may constitute sometimes an enslaving mental state when one cannot avoid thinking almost constantly about these things. To use a reflex to save ourselves from, say, an attack by someone, based on a quick impression is fine, but to ruminate day after day on this episode is enslaving. Many times, this prolonged over-elaboration of past events is very hard to overcome. Contemplative techniques help avoid this mental drifting, this quick elaboration and subsequent over-elaboration of sense impressions, so that we progress in our path to liberation.

Therefore, one immediate advantage of practising meditation is to improve the power of concentration. Meditation is a common Buddhist practice, something that today is becoming increasingly popular. We will not go into describing the several types of meditation, as our plan is not to describe all this in detail, and those interested in specifics can find a great number of texts on the topic (e.g. *Stages of*

Meditation, by the Dalai Lama, Snow Lion Publishing). But we will present the essence of meditation, the kernel of contemplative methods.

Because contemplative methods, call it meditation or otherwise, exist in many traditions, these techniques are not the sole property of Buddhism or Hinduism—of which these originated. Hence, Prophet Muhammad said, "One hour's meditation on the work of the Creator is better than seventy years of prayer". As well, the typical cloister of Christian monasteries where monks walk in silence serves its purpose as a place to meditate—in fact, there is a Buddhist practice called walking meditation. Indigenous people in North America have sweat lodge ceremonies. All of these traditions have a similar purpose. Following our habit to complicate matters, many types of contemplative methods have been created and named—call it yoga, sitting or walking meditation, mindfulness, contemplative prayer, etc. But once the essence of these diverse practices is distilled, it all comes down to just be able to quiet the mind in order to cultivate a personal capacity for deep concentration and insight. And the practice can be done everywhere, there is no real need for formal cross-legged sitting or any other posture that many times are part of the folklore surrounding these techniques (to be fair, there are reasons why some postures help the practice, like sitting straight, but the basic point is that even if you are confined to your bed, you can still contemplate). We should not become lost in the methodology. You can practise meditation while doing the dishes; just fully concentrate and observe those cleaning actions, the bubbling soap, and your hands becoming wet. Do not become concerned with anything else, just observe the, admittedly not too enthralling, dishwashing procedures. Don't worry, the problems of life will remain there waiting for you when you finish your "dishwashing meditation", and when you are done with the dishes, try to continue avoiding too much elaboration on thoughts, just apply the needed, minimal if possible, consideration of the future and past hardships. One reason to practise contemplative methods is that if some insight is obtained, then this realisation can alter what we tend to think of as stable mental traits, especially those that make us unhappy like anxiety or anger. It is said that at the start you do meditation, and later on, when you have become proficient in this, meditation does you, because the key point is not to be for just 30 or 60 minutes a day in that state of, let us call it, bliss or calmness. The point is to continue having the same mind state for the whole day.

And what is the notion of the self in Buddhist tradition? The self is conceived as an ongoing process, not as a stable entity but rather as a mirage of the mind: the experience of being someone results from the contents of the currently active self-model, which comprises an integrated model of the world and a self within it, these models being continuously generated. Human identity thus consists of an ongoing flow of psycho-physiological events. If you remember what we saw in Part I about how brains make models of the world around us and about our relation to it, then it is clear the similarity in the concepts between what is known today in the neurosciences and this religious tradition. Because these relations between an individual and the surroundings contribute to self-awareness, it is fundamental to recognise those relations if we are to understand ourselves (and our selves!). Hence, one purpose of contemplation is to realise not so much the nature of the self, but the *relations* of the self with events/environment in general. The meditative practices geared towards cultivating mindfulness—a word that has become very trendy—are not typically intended for relaxation but more to gain a sense of familiarity with one's own mental landscape. The essence of Buddhist philosophy is really psychology and ethics.

The main reason why these traditions are connected to the central theme of this book, that of being and becoming, is the focus of Buddhism in the acceptance of transience, the realisation of impermanence. The Buddha probably said that "All things appear and disappear because of the concurrence of causes and conditions, nothing ever exists entirely alone, everything is in relation to everything else" (we say "probably" because in his time he and his followers did not leave many writings, the texts appeared after his death, hence one can never be sure of what really happened, and this can apply to all ancient traditions; this is no criticism, it is just the way it is). To some extent, this perspective on transience can help us transcend dualities, those that we have mentioned in many paragraphs of this book, artificial separations that more often obscure than help our understanding of the nature of reality. So, reflecting on the personal existence, there is no substantially independent existent person, rather an assemblage of conditioned phenomena arising and passing away (in Buddhist technical words, this is called "dependent origination").

Translated to our neuroscientific parlance, this is what we explained in Part I about the models that brains make, and continuously update, of the personal identity and the surroundings. The physicists, who seem to agree too with this view, warn us: "What we find, therefore, are not elementary space-time realities, but rather a web of relationships in which no part can stand alone; every part derives its meaning and existence only from its place within the whole", words of Henry P. Stapp in his book chapter *Quantum Theory and the Physicist's Conception of Nature: Philosophical Implications of Bell's Theorem* (that appeared in *The World View of Contemporary Physics*, R. E. Kitchener, ed., State University of New York Press).

This is the most fundamental becoming we should realise and accept, the transience of things, of your body and self, the fact that both keep changing all the time, being updated as it were. This continuous change happens when you are eating a carrot and some of those carrot's nutrients will be part of your body soon, or it happens when you listen or read new information, and that information becomes part of your brain. But, when do these carrot or information pieces become you, your body? Is it when the carrot pieces are in your mouth or when its carbohydrate components are in your cells? And when you expel the remnants (breathing, urinating, defecating), are these compounds not you anymore? A famous scientist is reported to have said that today you are yesterday's potatoes. And as brains are concerned, today you are yesterday's thoughts and perceptions. When does a new piece of information become you? Possibly when it is incorporated in your neural circuits. But if you were to suffer amnesia and cannot access this knowledge anymore, are these information bits still part of you? Things to ponder on a sleepless night.

Anyhow, both contemplative traditions and scientific findings can help us comprehend the nature of our psyches and our lives, which is a start in the path of liberation from the conditions which give rise to suffering. Any help should be welcome. Especially, help with dealing with our egos. Let us see whether these egos can be diminished.

Chapter 15
Shrinking the Self

In Chap. 5 we saw that eliminating the self is almost impossible; nonetheless, it can be shrunk, diminished to some extent so that one can be somewhat emancipated from the chains imposed by the sense of a big personal identity. This is another basic tenet of Buddhism, that of attachment/non-attachment. Phenomenal self-consciousness involves a relation between the self and objects; the more and stronger the connections between the individual and objects or other people, the greater nourishment for the sense of self. In our world with so many opportunities to possess things—and people too—, being so easy to develop cravings, robust attachments arise. An ordinary example that is the subject of countless books, movies and songs is infidelity in romantic relationships. Rooted in the self, in the attachments of that self to the partner and the desire for possession, jealousy emerges if a partner has a physical, sexual or emotional relationship with another person. Interestingly, if the other relationship, instead of being physical or emotional, is intellectual, it may be much less painful—there seems to be a different perspective if the intercourse is sexual or intellectual. Have you ever experienced any of this? Did you feel a difference if your partner was involved sexually, emotionally or intellectually with someone else? Why is that? Please elaborate on this (but do not over-elaborate!) before reading the next paragraph.

If that happened to you, you were experiencing the drives/power from your genes, the result of the already mentioned in Part II forces of nature. The degree of possession—in part driven by the genes—is higher in a love relationship than in a friendship, and for this reason, when the connection, the romantic bond, may be falling apart the self becomes distressed. The stronger the attachment, the more probable the disappointment and the grief. Genes do not care much about philosophising, they care about reproduction, so they have engineered—so to speak figuratively of course because genes, once again let us repeat, do not know what they do— neural circuitries in your brain to predispose you to be extremely possessive about the person with whom you may pass on your genetic material. Being just friends there is not much afterthought about where and with whom your friend has

J. L. Perez Velazquez, V. Nenadovic, *Being and becoming*,
https://doi.org/10.1007/978-3-030-78264-1_15

been, but in a romantic relationship, there is tribulation about these matters because the neural networks sculpted by the genes are fully working on making sure you, as a mammal, will receive the sperm (if you are female) or you will exclusively enjoy the ovum (if you are male) of your partner. And you thought you were jealous! For those interested, the biology and neuroscience of attachment (also known among humans as love) is a most fascinating subject, with a very large number of studies reporting findings obtained in animals and humans that shed light on what really love is all about.

The point is that once the seeds of the self are planted, its growth is inevitable. Our Western society, especially, is a culture medium for the development of selfhood, in the early stages of life fostered by parents, friends, teachers, and late in life colleagues and other events that will make the individual believe there is something special about him or her. There is more and more weight in our times about the need to nurture that sense of identity, to convince the child, the individual, that he/she is a most remarkable natural wonder. Context may help to cast doubt on this emphasis: in our current global population of 7.8 billion people, how special can one person be? It was not the same, by the way, in the times of my youth. In fact, the situation was almost the opposite: teachers would not hesitate to remind us how silly and ignorant we were and would fail us in exams without hesitation; today, failing a student has to be considered carefully, lest his self-esteem becomes hurt. We are not going to repeat again the multiple drawbacks of having a big ego. There should be some balance between reminding one of his ineptness and praising every act he performs.

What the Buddha prescribed for dealing with ever-increasing selfhood is the consideration of transience and non-attachment. Please note that non-attachment has a slightly different connotation from detachment. The latter normally denotes disinterest, indifference; if one is detached from the pets at home, it means he/she does not care too much for them. But if one is non-attached, you can still care a great deal and try to make them have a happy life, but at the same time you understand and accept the fact they will be gone, probably, before you, and that there is little point in suffering too much grief because you are already expecting this outcome. Because, let us be honest, when people feel sorrow for the loss of someone, it is mainly because they will miss the departed, because they would like to still have them in our lives. So there is some self-interest lurking here. Living with people, our brains create a world with these people and our attachments to them. Much like the phenomenon of phantom limb pain (described in Sect. 1.8), the circuitry and networks representing the loved ones remain; the feedback that would have existed in our interaction with them is gone when they die, and thus pain—this time psychological rather than the physical of the phantom limb—ensues. This is particularly poignant in parents who have experienced the loss of a child. In our current Western society where infant and child mortalities are low, there is the expectation that our children will both grow into adulthood and that they will outlive us. If you are middle-aged, you may have discovered that this was not the expectation of your grandmother's generation, and it is also not something taken for granted in other parts of the world. However, people of all ages die, and realising the transient nature of everything, of our relations and our possessions, will help us when the time comes to lose them. This is, in a nutshell, the essence of the teaching of non-attachment.

As an aside, about this business of self-interest, it is of note that all actions, even those considered unselfish, empathic, and disinterested, contain some element of self-interest, or self-centredness. A saint, a holy person, performs his or her good actions due to the fact that this makes him/her happy, so one can say that in a selfish manner the saint is performing an (apparently) unselfish action, and that's the way it is. Remember that in Sects. 1.3 and 2.2.2, it was described how our brains are attuned to rewards, the many neural pathways all over the brain that process pleasure and reward, for this is the foundation of animal and human behaviour: the avoidance of punishment and the pursuit of reward. So do not be embarrassed to realise that you behave in the way you do because this satisfies you; rather you should be ashamed of doing things that make you unhappy. But naturally, our willingness to admit to our motivations is shaped by family, peers, society and our specific cultural and religious teachings, and these external modifiers influence our primary reward circuitry and add a moral dimension.

If our attachments enlarge the ego, with all the associated problems already mentioned in Part I (vulnerability, distress, etc.), then perhaps reducing or eliminating the attachments will help diminish the ego/self. Time for another thought experiment. In Fig. 15.1, there is an individual who has established many relations, attachments (the arrows in the cartoon) to things and people around. She possesses material things and people too, for she is madly in love and desires that her partner belongs only to her. She has cravings of all sorts. In the second panel of the figure, she has managed to get rid of some of those arrows, and she has reduced or lost some attachments; she is eliminating little by little some dispositions like fear and aversion, or some cravings, or even the judgments about liking and disliking. Then, as a result, her ego/self will be diminished; she will be less vulnerable to insults and criticisms, expectations not achieved, psychological injury, and even to physical injury, for now that she has vanquished aversion and liking and disliking thoughts she understands and accepts that pain—which is not to like or dislike—is there to signal that something is wrong and hence you can move forward to try to find a remedy. In short, she experiences some advantages in this thought experiment about practising non-attachment; she benefits from eliminating desires and cravings, those fetters that many times suffocate us—yet other times give us pleasure. One wonders whether in the end, if she gets rid of absolutely all relations/attachments, her sense of self will become so tiny that may be unnoticeable. Those of you who have dealt with Buddhist practitioners—the serious ones—may have noticed that their sense of pride, their egos or selves, are very much imperceptible.

Naturally, to eliminate attachments is easier said than done. It is not that we are advocating embracing pure and strict stoicism; the point is to reach that equilibrium mentioned in the previous section, neither too little nor too much, just the right amount. You can fully enjoy the company of your friends or a bottle of wine,

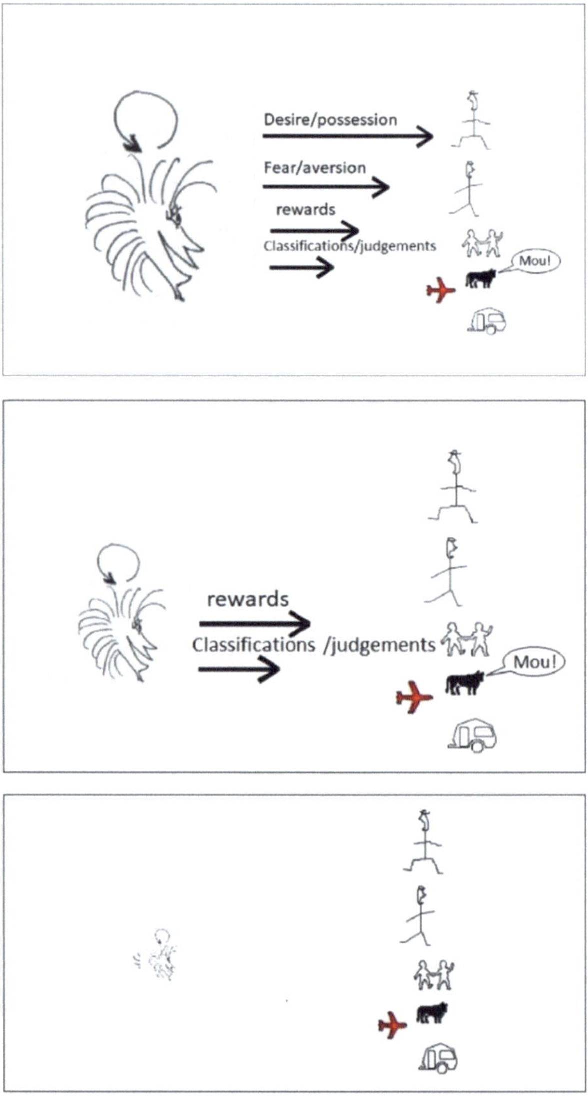

Fig. 15.1 Self-enhancement of the self. The individual on the left has many relations, traits and strong attachments (depicted by arrows) to the things/people in the right hand side. She desires possessions, craves rewards, hates and loves things and people, likes and dislikes, and judges and prejudges. These relations contribute to enhance her sense of self, in a sort of closed loop within her mind. As she is able to eliminate these arrows, these fetters of her life, these strong attachments, her sense of self —her ego— diminishes, as explained in the text

but realising and accepting that everything changes, that the "becoming" is there all the time, and that sometime in the future those friends or that wine will not be there anymore. Minimise attachments, and you minimise sorrow, and especially the attachments to ideas are perhaps the ones that most need to be reduced—just look at some world events and witness the power of attachments to beliefs and ideas/ideals.

Chapter 16
Naturalising Death: The Ultimate Becoming

All life forms long to persist in their being. Perhaps it is the notion of nonexistence that is so frightening, unfathomable—at least for those animals that can consider nothingness, and we may not be the only ones, see below. At the same time, it is striking that people seem scared to think about nonexistence after death, but not prior to birth. Why is posthumous nonexistence more frightful? Should I be frightened to know I was not present about one hundred years ago to witness the armistice after the first world war, the war to end all wars as they said? Should my hairs stand on end when I realise I was not around 2000 years ago when the roman legions of Octavius Augustus were building my hometown, Zaragoza, in Northern Spain? If not, why should I be scared on account of the fact I will not be here 100 years from now to witness what has become of my favourite football team? The reflections of Emperor Marcus Aurelius should not be forgotten: "It is not death that a man should fear, rather he should fear never beginning to live".

The last act in our chain of being and becoming is death, although not really because, while our personal identity is dissolved after losing mind and body, our bodies become something else. Death could thus be considered the final becoming for us. As the nervous system activity dwindles and its cell networks cannot maintain the model of the self, upon the total cessation of brain activity, our selves disintegrate and evanesce, while our biochemical components and the body energy become something else.

It is not only humans that fear this final moment; just watch any animal and see how fast they avoid injury and potential demise. Albeit many animals will not know about death, they just follow the reflex-like behaviour imprinted in their nervous systems—one wonders whether some primates or other brainy animals have some concept of death, and to see how one can go about investigating this issue, you can read Monsó's paper "How to Tell If Animals Can Understand Death" (Monsó,

J. L. Perez Velazquez, V. Nenadovic, *Being and becoming*,
https://doi.org/10.1007/978-3-030-78264-1_16

2019), and also there is some indication that some animals do have a sense of passing away as can be inspected in a brief report of the National Wildlife Federation.[1]

In any case, the fear of death or injury is understood once we consider that genes created our bodies to be generating more genes; hence, they took special care to have brain networks attuned to the possible perils out there and to make sure the organisms run away from these so that they can proceed to spread the genetic materials. People in different cultures throughout history have devised scenarios to avoid death or at least eternal nonexistence. The most popular is to postulate the existence of an afterlife, normally involving a heaven, hell, or purgatory. Since nobody has ever come from those places, I guess we will have to wait until our time comes to see whether or not this is real (this will be our last experiment, conducted by scientists and laypeople alike). Another strategy to avoid extinction is reincarnation. There are some logical problems, though, with this notion, starting with the fact that there have been a huge number of people on the planet, more and more as time passes, so there seems to be a surplus of souls of previously deceased individuals over living bodies; also, one wonders who was the first one to reincarnate before there were humans on Earth. Of course, we can go on and postulate things that sound more adventurous every time, like prehuman animals developing souls or that the many people who have lived take turns to reincarnate in the limited number of living bodies at any time. In the end, this line of reasoning becomes a pile of hypotheses on top of hypotheses. It does not help much.

The plea to escape the void has been, in part, the reason for the rise of religions. Another reason which inspires popular and religious traditions and beliefs is the vicissitudes and ordeals of life: there is the danger that hailstorms will destroy the crops, hence better to create a deity protecting the crops. Things have not really changed too much in this business during these past many centuries, only the names. If in the past some gods were protecting against hail, in modern times these are saints (at least in the Christian tradition), the patron against hailstorms being St. Barnabas. And for the sake of completion, let us mention the third reason for religions to emerge, the awe ancient humans felt about the wonders of nature and the associated psychological need to find an agent causing these marvels (the famous agency so much discussed in part I); think of the impression a troglodyte would have of watching a rainbow appear and disappear—some readers may know that in Judeo-Christian tradition, the rainbow is considered a covenant of humanity with God. Nevertheless, this text does not intend to tell people what they should believe or not, so let us stop these considerations, although some are quite amusing.

One thought that may be running through some readers' minds now is that in this writing we are insisting on the "becoming", that things do not disappear rather become something else—and physicists know this well because when elementary particles vanish, they become some other particle or energy—, then how come consciousness, or the self, disappear. But consider another thing we have as well maintained in this book—that these are concepts (consciousness, life…) we have created

[1] www.nwf.org/Magazines/National-Wildlife/2013/DecJan/Animals/Animal-Mourning.

to capture a constellation of phenomena, so the atoms that constitute you right now will become something else tomorrow and when you die, but the general notion of your life and your consciousness will be gone. *It is the concepts that disappear, not the fundamental constituents*. To quote a law of thermodynamics, "The total amount of energy and matter in the universe remains constant, merely changing from one form to another". It can be a comforting thought that your fundamental constituents will live forever, but not your sense of personal identity.

This final, inevitable dissolution of the self was once part of our shared experience and shaped our views of death and dying. In the not-so-distant past, people of all ages died at home and were temporarily laid out in the house for the gathering of mourners prior to burial. For centuries before this, imprinted on our neural circuitry was the representation of the cessation of heartbeat and respiration and the ensuing decomposition. The rise of hospital care with modern critical care and life-sustaining technology has changed this. Death now is largely hidden from our everyday lives, and its definition has changed (the several definitions of death were introduced in Sect. 2.1.1). Sustaining respiration with ventilators and circulation with multiple medications or even extracorporeal membrane oxygenation (heart-lung bypass) has made physical death difficult; while the self is unconscious and the person has lost any meaningful interaction with us, we will still see a chest rise and fall and an electrocardiograph display each heartbeat. The last ventriloquist has been silenced, and when technology is removed, the heart and lungs will shortly cease to function.

Chapter 17
Law and Neuroscience: The Impact of Brain Research on Criminal Justice

Contents

In a book like this where we are considering how neuroscientific findings are enlightening some aspects of our lives, we should not forget to consider another important facet, the law. Neuroscience research is having an impact on the legal system, especially criminal justice. As promised in Chap. 2, we will very briefly comment on this point. And if some readers are into science fiction, they can read a couple of stories about the control of pathological brain activity using neurostimulation—more specifically deep brain stimulation—that went badly, where the issue of legal responsibility for crimes committed by patients under treatment is developed: *DBS—An odyssey through the Mind*, and if you prefer reading Spanish, a similar version is *La lectora de Sueños* (both of these science fiction stories are based on work performed in our laboratory on the control of brain activity using deep brain stimulation).

The advances in brain research have driven the reflection on the repercussion the new knowledge may have on the understanding of the legal concepts of responsibility and morality. That this is becoming fashionable is heralded by the creation of such fields as neurocriminology or the rise of terms like neurolaw that, in the words of one author, is "an interdisciplinary field which links the brain to law, facilitates the pathway to better understanding of human behaviour in order to regulate it accurately through incorporating neuroscience achievements in legal studies" (Petoft, 2015). We should consider that laws are based on the notion of morality, an extremely arbitrary concept. Morality emerges from the self or selves in a social environment or community. Morality does not exist in nature—remember M. Gazzaniga's words in Sect. 2.2.2 about "Personal responsibility is a public concept. It exists in a group, not in an individual".

It is of interest that in times not too long ago—mid twentieth century—scientists talked about "cerebral dysrhythmias" as the cause of neuropsychiatric disorders,

J. L. Perez Velazquez, V. Nenadovic, *Being and becoming*,
https://doi.org/10.1007/978-3-030-78264-1_17

especially aggressive behaviours. In spite of the primordial state of neurophysiology in those early ages of brain recordings, already some scholars called for changes in the legal system associated with criminal behaviours. And despite the current plethora of sophisticated methods for recording and analysis of neurophysiological activity, the results do not provide yet a clear picture of the brain dynamics in this type of psychiatric patients. Nonetheless, putting together the outcome of many observations on structural and functional brain studies, a picture emerges indicating that, while details are not completely known, the brain dynamics of these patients who commit crimes, like serial offenders, are different from those of the healthy population. The subjects afflicted with these disorders have a diseased brain, and the general question for the criminal justice system becomes whether or not these individuals can be held accountable for their actions and to what degree. Recent thoughts on the subject can be found in the paper "Neuroscience Has the Power to Change the Criminal Justice System" (Altimus, 2017), or in the DANA Foundation report "Will Neuroscience Challenge the Legal Concept of Criminal Responsibility?" by K. Sukel.

It is impractical to even try to summarise here the data on neurophysiology and neuroimaging done in criminals. The few things that can be clarified now is that there have been found structural and functional abnormalities in the brains of serial killers and other serial offenders, which are supposed to be a reason why their actions represent an antisocial variant of a normal animal behaviour—called phylogenetic predation—but that has gone out of context, as explained in Chap. 10.

Today, scientists and clinicians are talking about neuropsychiatric phenotypes in many psychiatric deviations (phenotype refers to the multiple characteristics observable of an organism such as morphology and behaviour, whereas genotype refers to genetic material). A conclusion derived from studies conducted by the FBI indicated that a fundamental characteristic of these psychiatric syndromes (addiction, psychopathy and other syndromes like anorexia nervosa) is that they start early in life and remain very stable states throughout the individual's life. The predicament of these patients/criminals is that their brain abnormalities sometimes prevent them from curbing their actions. As to why some are not endowed with the willpower to behave in a socially acceptable manner or to overcome addiction or whatever syndrome they may have, well, recall the aforesaid words: their brains are ill. It was stated in Sect. 2.2.2 that brain diseases leading to deviant behaviours seem to be given a different perspective among people than those of other organs like kidneys because brains determine the conduct we observe in people, whereas kidneys do not (except when they fail to clear toxins that affect the brain). When that conduct is anomalous, is antisocial, or falls outside the relatively safe central part of the Gaussian curve of socially accepted human behaviours, it seems an almost general consensus that the person afflicted with that anomaly should be able to, upon reflection on those deeds, change his/her actions. In other words, the person should be able to control his/her own brain and behaviour. But we have already seen in Part I that this notion of a central command inside the mind with logical powers enough to overcome the rest of brain circuitries—that last ventriloquist of Sect. 2.2.1—is to a large extent a fallacy, and this has important implications in societal and legal aspects. A brief note: for those unfamiliar with the aforementioned Gaussian curve, let us just say that it is a (bell-shaped) representation of a probability distribution that is symmetric about the mean, showing that data near the mean are more frequent in occurrence than data far from the mean.

With our advanced neuroscientific methods, some readers may be wondering how one can be sure that the actions that lead to a criminal conviction are due to brain damage. The short answer is that we cannot be sure, at least at this point in time, as there are many brain things we do not know. Nevertheless, the neuroscientific findings in the nervous systems of criminals are enough reason not to be too judgemental. That's not to say criminals have to be left free wandering in the society; dangerous people should be kept away, but where they can be sent and how they can be put to do something useful is beyond the scope of neuroscience... Or perhaps not completely, as science still can inform as to who may be "fixed" and who may be a lost case.

Legal systems, especially laws involved in criminal actions, have to contend with more or less accepted notions of sanity and insanity, and thus, there is an important contribution of medicine and neuroscience. We have proposed that the self and its actions emerge from brain activity and that brain circuitry responds to its environment and produces behaviour. The self does not emerge in isolation but in constant reciprocal interaction with its environment. The criminal justice system too emerged from interactions: from societal laws and norms dictated by other selves' brains, typically those holding social and political power. These particular selves have created the societal network that dictates appropriate societal behaviour. On November 25, 2016, the Government of Canada introduced legislation to eliminate Sect. 159 of the *Criminal Code*, which makes anal intercourse illegal except when it is conducted in private between two persons who are both 18 years of age or older or between a married husband and wife. Would we classify anal intercourse as an act driven by a diseased brain, instead of a search for pleasure? Criminality is often linked to psychiatric disorders (diseased brains), and definitions of what constitutes psychiatric disorders change over time. Homosexuality was once considered a psychiatric disorder in need of treatment. Whereas psychopathy is characterised as antisocial behaviour—among other aspects like impaired empathy—it is not without interest that research has shown that traits that fall in the category of psychopathy may have an evolutionary advantage. It may be that psychopathic traits enable that self to take more risks and be less attached to people, thus engaging in sexual activity earlier and with multiple partners, spreading their genes which ultimately dictate their behaviour without the benefit of morality. Curious readers can consult texts like *The Wisdom of Psychopaths* (Dutton, 2013) to learn about this intriguing condition and the fact that psychopaths are all around us.

We will finalise these deliberations with something perhaps not many people would like to hear: if we cannot be too harsh to judge others in their bad behaviours because their brains are sick, the same should apply to good behaviours. Good behaviour may not be a product of an ill brain, but there is always some self-interest lurking in our best, unselfish and altruistic behaviours (already talked about in Chap. 15). So in the final analysis, either good or bad actions, the roots are in the neural circuits that populate our brains and bodies. Attaching a moral judgement to these behaviours in terms of good or bad will not change the fact that this person behaves like this because he has this brain, and not another.

17.1 Postscript to Part III

In this part, and in fact in the previous two as well, selfishness, self-interest and related psychological features have appeared prominently. It may be thus of interest to contemplate the possibility that our modern societies are rooted in the self, on egotism. It is known that today's societies are mostly hierarchical, stratified, while in the very distant past (that of the hunter-gatherers), societies were mostly egalitarian. The transition—the becoming—from egalitarian to stratified societies was advanced from the moment things appeared that could be possessed and amassed, mainly after the invention of agriculture. Human societies became stratified in the sense that hierarchies were established; inequality thus emerged, in the distribution of goods, money, resources, etc. But why is it that egalitarian cultures basically vanished? One reason is that egalitarian societies are unstable. A society based on traits almost everybody share will be more stable than another based on personality features that are not too common in the whole population. What is the most common trait among humans? Selfishness, self-centredness, craving for possessions. The tendency toward selfishness has been a reason to explain the success of stratified cultures.

Why these comments on very basic aspects of societies are mentioned within the context of the dynamics of life and consciousness? It is precisely because this represents an illustration about the dynamics of our cognitive development (in terms of selfhood mainly) and life, in this case the making of cultures and our artificial ecosystems we call societies—sociodynamics, or social dynamics, is the field that studies the behaviour of groups resulting from interactions of individuals. For those interested in details about how unequal societies came to displace egalitarian cultural norms over time, please consult the paper "The Spread of Inequality" (Rogers et al., 2011) where a quantitative model is presented that incorporates several variables like mortality and fertility rates, resource availability, etc. In short, because traits like egoism and self-centredness—natural tendencies fostered by the genes after all—are ubiquitous among all humans, a culture/society based on these traits will be more stable than those based on empathy, generosity, altruism, and the like. This may be a reason too why communism/socialism has failed in many societies, as the fall of the Berlin wall illustrated, for to be a stable communist culture requires most people to be generous enough to be able to go along without much private property, but, who does not love private possessions. We are just not ready yet. At the same time, it should be considered what the biologist and writer Edward O. Wilson said, "Selfishness beats altruism within groups. Altruistic groups beat selfish groups". There you go, a succinct explanation of our societies (those groupings we make, based in part on prosociality and on shared perceptions, on collective agreements about ideologies and beliefs, yoked by more material arbitrary markers such as customs, language accents, garments, etc.) and of the success of Homo Sapiens over other animal and human species.

Whereas it may be a bit disturbing to some that selfishness, egoism and similar traits are at the roots of the sociodynamics from where modern societies emerged, the realisation of this fact represents the possibility of change. Recall Chap. 10 comments on the intellectual prowess afforded by our brains and how this intellect could overcome the natural tendencies related to self-interest. It is all in our minds.

Part III has shown how some philosophical and religious traditions share many points of view with the findings of science; in fact, some ideas were anticipated long before science came into place to indicate, or even demonstrate, the validity of those proposals. Specifically, the acceptance of the becoming of things, events, and of our sense of personal identity as well, is crucial to avoid becoming lost in today's plugged-in world, and these acceptances are tenets of the Buddhist tradition.

Because in reality, life is simple, but we make it complicated. The bare necessities of life, finding shelter and food, are within most people's reach (mind you, there are still people around the globe that cannot find shelter or proper nourishment, but we are talking in general). This is not to say life was always simple. Just ask any hunter-gatherer of 15,000 years ago, and he/she will tell you the predicaments of the group to find food, water, shelter, and sexual mates. He/she would be equally impressed when told about our modern times, where all these needs are at our hands, impressed not so much because of this fact but because we have created such a large number of other "bare necessities" and the associated frustrations when these are not met, that maybe our hunter-gatherer will be left with the impression that it is better to go back to his hunting and gatherings than being immersed in a mostly artificial and bewildering world where the satisfaction of one need leads to the craving of another. In a world with uninterrupted noise, with a cognitive space overcharged with incentives to act, to possess things, where antidepressants are consumed to meet the unending pressure of production, these are aspects contributing to the alienation of our times. But we do not have to go back millennia to appreciate a different and more balanced worldview. In Thomas King's short story *The One About Coyote Going West*, a North American indigenous worldview is presented. Coyote, the mystical trickster, representing the Indigenous worldview, is admonishing "the mistake"—the White man she (Coyote) created—which is producing all manner of material things from a catalogue. She tells the "mistake" to stop: "*We don't need that stuff.... You got to stop making all those things. You're going to fill up the world.... You got to give me that book before the world gets lopsided*" (p. 208); there is the implication that the world will be off-balance with all of the complicating things made by Coyote's creation that indigenous people did not need.

If instead of being tethered to cell phones and over-elaborating on events we were to practise equanimity, to calm our minds of that never-ending mental wandering of thoughts, the beginning of liberation would emerge. And as explained in this book, once a pattern emerges, its sustainability depends on so many factors that its continuation is never guaranteed; yet there may come a point when the pattern becomes so stable, when a sturdy equilibrium is reached, that it will go on for a long, long time.

Chapter 18
Final Conclusions

In our journey through the scientific discoveries and philosophical traditions described in the three parts of this book, we have seen that consciousness and life are concepts that we have created to encapsulate a constellation of phenomena, from self-awareness to agency and all the features associated with living processes. Particularly, we have seen that the emergence of selves is an inevitable consequence of embodied brains immersed in an environment, arising from the primordial necessity to situate the organism in the environment and developing through the perceptions of our behaviours within those surroundings. We have understood too that a self represents some evolutionary advantages for the entity, but we have learned as well that these selves are models, illusions to some extent, created by that ultimate illusionist that is the brain. And we have also realised the perils of a strong attachment to that model, hazards that extend beyond the psychological vulnerability spreading towards the need to find reasons to live, a meaning in life, life that we have seen was another inevitable consequence due to the chemistry of organic compounds and the physics surrounding them in a planet like Earth. We have thus understood that there is nothing really extraordinary about the emergence of life and us within nature. Rather, what is special about humans is that the possibility has appeared for the first time in nature, thanks to the rise of highly evolved intellects, to go against the forces of the genes, to perhaps enter a new era where it is not the transmission of genetic material but the transmission of mental symbols, ideas, and cognition that will determine what humanity will become.

Being and becoming is the essence of reality, something everybody knows but few truly comprehend. Contemplating this, meditating on these most basic aspects of reality, is the start in the path of liberation from the chains of our created life. It has been a contention of this text that the comprehension of the teachings of neuroscience and physics on the transient nature of all natural phenomena—included us—will aid that realisation of impermanence. This is not to say that everyone should embrace science as guiding our every waking moment, although in truth we are

J. L. Perez Velazquez, V. Nenadovic, *Being and becoming*,
https://doi.org/10.1007/978-3-030-78264-1_18

completely welded to the technological aspects of science, but that a basic understanding of experimental observations and scientific theories help that undertaking.

The take-home message of these scientific findings and musings about our nature may have been disappointing to some readers. There is no need to repeat the reasons, expounded in the postscript to Part II, on the perchance unsettling sensations about the lack of deep meaning of life, about the results of our anthropocentric tendency to think in terms of agents, causes of things and processes, and why these disillusioning feelings need not be such. It is thus conceivable that the lack of very clear and specific causes for the emergence of some phenomena treated in this book, like life and self-awareness, could be somewhat unsatisfactory. In our descriptions, it has become apparent that establishing a definite cause for a complex phenomenon is not only impractical but also unfeasible. In some other circumstances, in simpler systems, it may be possible to discern a main cause for the event, say for the appearance of lightning (a specific cause is the distribution, the imbalance, of electric charges in the clouds). But if the study of complex systems teaches us something, it is that reasoning in terms of strict cause-effect relations in these complex, nonlinear systems where there are interactions at multiple levels among the system constituents is difficult if not impossible. Even in apparently simple situations like the Bénard convection of Sect. 7.1, it was hard to point out the main cause of the phenomenon. But this state of affairs when describing complex systems and basically the things important to us like our sense of personal identity or the "meaning" of life need not be unsatisfactory. On the contrary, it provides a more holistic perspective through which the global nature of phenomena can be envisaged. It can afford us an intuitive sense of how nature works, and this may be more than enough for some of us.

Included in this intuitive understanding is that everything changes (recall Suzuki Roshi's words explaining Buddhism). The perspective of time, or how the passing of time alters the perspective, may be central for an additional push towards liberation from the consequences of our continuous ruminations and over-elaborations on perceptions and thoughts. Today you hate someone, and there is the chance that in the distant future, you will look back on this hatred and find it odd. One mental exercise I perform in cases when a heavy burden afflicts me consists in thinking about how I will see this event in the distant future, because as life has taught me, time cures (almost) everything, and what yesterday was an unsurmountable agony today looks like a trivial affair. This exercise relieves me; perhaps, it may help others. And if that is not enough, there is always refuge in thought, for, as Horace Walpole told us, "Life is a comedy to those who think and a tragedy for those who feel"; in the theatre of existence, serene reflection is always of great service.

If the insight of becoming may be important, the understanding of our possible roles in these becomings is also crucial for what will, in the end, come to pass, because we are able to understand all these things due to our intellect. Such is our lot, that nature, via the genes and other circumstances, has endowed us with material to create the nicest of arts and the most devastating forces our planet has seen (only rivalled by the Permian extinction that wiped out about 95% of living forms). As species go extinct at unprecedented rates, as individuals become attached to ideologies and beliefs, as we lose ourselves in the overabundance of possibilities to consume where one craving follows another, it is worthwhile to apply those cognitive powers. We can foster education of the youth, such that if the central theme of medieval thought—and other past times—was sin and eternal damnation versus everlasting salvation and glory in the afterlife, in future times it will be *the fulfilment of the present experience of our understood realities* (and this sentence condenses what we think about the meaning of life). And when the time comes, when parts of us will become something else, we can conclude that it was a life worth living. Although we can feel constantly that self creeping, sneaking among our thoughts, dictating a reality in front of us, other ventriloquists in our minds will cast doubt on our perception of that reality and will recognise that what we are today may not be what we will be tomorrow, fully realising too that next moment may be totally different. Because everything changes.

As a concise encapsulation of the matters treated in this text, it is reasonable to end with these words written not by a scientist or philosopher, but by an observer of people (English translation on the right-hand side):

… y en el mundo, en conclusión,	… and in all the world, in conclusion
todos sueñan lo que son,	Man dreams whatever he be,
aunque ninguno lo entiende.	Even though nobody understands it.
[...]	[...]
¿Qué es la vida? Un frenesí,	What is life? A frenzy,
¿Qué es la vida? Una ilusión,	What is life? An illusion,
una sombra, una ficción,	A shadow, a fiction,
y el mayor bien es pequeño;	And the greatest good is but small,
que toda la vida es sueño,	That all life is a dream,
y los sueños, sueños son	And dreams are only dreams

Pedro Calderon de la Barca, *La vida es sueño* (Life is a Dream), 1636

References

Adamatzky, A., & Akl, S. G. (2011). Trans-Canada Slimeways: Slime mould imitates the Canadian transport network. *International Journal of Natural Computing Research (IJNCR), 2*, 31–46. *arXiv*:1105.5084v1.

Altimus, C. M. (2017). Neuroscience has the power to change the criminal justice system. *eNeuro, 4*(1), e0362-16.2016.

Ambrose, E. J. (1982). *The nature and origin of the biological world.* Ellis Horwood.

Ariely, D. (2008). *Predictably irrational: The hidden forces that shape our decisions.* Harper Collins.

Ashby, W. R. (1962). Principles of the self-organizing system. In H. von Foerster & G. W. Zopf (Eds.), *Principles of self-organization: Transactions of the University of Illinois Symposium* (pp. 255–278). Pergamon Press.

Assal, F., Schwartz, S., & Vuilleumier, P. (2007). Moving with or without will: Functional neural correlates of alien hand syndrome. *Annals of Neurology, 62*(3), 301–306. https://doi.org/10.1002/ana.21173

Bak, P., Tang, C., & Wiesenfeld, K. (1988). Self-organized criticality. *Physical Review A, 38*, 364–374.

Baracchi, D., Lihoreau, M., & Giurfa, M. (2017). Do insects have emotions? Some insights from bumble bees. *Frontiers in Behavioral Neuroscience, 11*, 157.

Bértolo, H. (2005). Visual imagery without visual perception? *Psicológica, 26*(1), 173–188.

Blanke, O., & Arzy, S. (2005). The out-of-body experience: Disturbed self-processing at the temporo-parietal junction. *The Neuroscientist., 11*, 16–24.

Blanke, O., & Mohr, C. (2005). Out-of-body experience, heautoscopy, and autoscopic hallucination of neurological origin—Implications for neurocognitive mechanisms of corporeal awareness and self-consciousness. *Brain Research Reviews, 50*(1), 184–199.

Blanke, O., & Thut, G. (2007). Inducing out-of-body experiences. In *Tall tales about mind and brain* (pp. 425–439). Oxford University Press.

Bohr, N. (1934). *Atomic theory and the description of nature.* Cambridge University Press.

Boyan, G. S., & Reichert, H. (2011). Mechanisms for complexity in the brain: Generating the insect central complex. *Trends in Neurosciences, 34*, 247–257.

Bull, M. J., & Plummer, N. T. (2014). Part 1: The human gut microbiome in health and disease. *Integrative Medicine (Encinitas), 13*(6), 17–22.

Cairó, O. (2011). External measures of cognition. *Frontiers in Human Neuroscience, 5*, 108. https://doi.org/10.3389/fnhum.2011.00108

Callender, C. (2001). Taking thermodynamics too seriously. *Studies in History and Philosophy of Modern Physics, 32*, 539–553.

J. L. Perez Velazquez, V. Nenadovic, *Being and becoming*,
https://doi.org/10.1007/978-3-030-78264-1

Cammaerts Tricot, M. C., & Cammaerts, R. (2015). Are ants (Hymenoptera, Formicidae) capable of self-recognition? *Journal of Science, 5*(7), 521–532.

Castro, A., Díaz, F., & van Boxtel, G. J. M. (2005). What happens to the readiness potential when the movement is not executed? *NeuroReport, 16*, 1609–1613. https://doi.org/10.1097/01.wnr.0000183331.51736.7f

Cohen, M. A., & Dennett, D. C. (2011). Consciousness cannot be separated from function. *Trends in Cognitive Sciences, 15*, 358–364.

Craig, A. D. (2003). Interoception: The sense of the physiological condition of the body. *Current Opinion in Neurobiology, 13*, 500–505.

Damasio, A. R. (1999). *The feeling of what happens: Body and emotion in the making of consciousness*. Harcourt Brace.

De Ridder, D., Van Laere, K., Dupont, P., Menovsky, T., & Van de Heyning, P. (2007). Visualizing out-of-body experience in the brain. *The New England Journal of Medicine, 357*, 1829–1833. https://doi.org/10.1056/NEJMoa070010

Desmurget, M., & Sirigu, A. (2009). A parietal-premotor network for movement intention and motor awareness. *Trends in Cognitive Sciences, 13*, 411–419.

Dickinson, W. J., & Seger, J. (1999). Cause and effect in evolution. *Nature, 399*, 30. https://doi.org/10.1038/19894

Droege, P., & Braithwaite, V. A. (2015). A framework for investigating animal consciousness. *Current Topics in Behavioral Neurosciences, 19*, 79–98.

Dutton, K. (2013). *The wisdom of psychopaths: What saints, spies, and serial killers can teach us about success*. Anchor Canada.

Dyson, F. (1999). *Origins of life* (2nd ed.). Cambridge University Press.

Edelman, G. (2001). Consciousness: The remembered present. *Annals of the New York Academy of Sciences, 929*(1), 111–122.

Eigen, M., & Schuster, P. (1982). Stages of emerging life—Five principles of early organization. *Journal of Molecular Evolution, 19*, 47–61.

Feinberg, I. (1978). Efference copy and corollary discharge: Implications for thinking and its disorders. *Schizophrenia Bulletin, 4*, 636–640.

Ferguson, R. B. (2018). Why we fight. *Scientific American, 319*(3), 76–81. https://doi.org/10.1038/scientificamerican0918-76

Frith, C. (2005). The self in action: Lessons from delusions of control. *Consciousness and Cognition, 14*(4), 752–770.

Gallagher, S. (2009). Neurophenomenology. In T. Bayne, A. Cleeremans, & P. Wilken (Eds.), *The Oxford companion to consciousness* (pp. 470–472). Oxford University Press.

Gallup, G. G., Jr. (1970). Chimpanzees: Self recognition. *Science, 167*, 86–87.

Gazzaniga, M. S. (1980). *Progress in brain research—Motivation, motor and sensory processes of the brain* (Vol. 54). Elsevier/North-Holland.

Gazzaniga, M. S. (2005). *The ethical brain*. Dana Press.

Gazzaniga, M. S., & LeDoux, J. (1978). *The integrated mind*. Plenum Press.

Gell-Mann, M. (2001). Consciousness, reduction, and emergence. *Annals of the New York Academy of Sciences USA, 929*, 41–49.

Gloor, P. (1986). Consciousness as a neurological concept in epileptology: A critical review. *Epilepsia, 27*(Suppl. 2), S14–S26.

Gnedenko, B. V., & Kolmogorov, A. N. (1954). *Limit distributions for sum of independent random variables*. Addison Wesley.

Goldenfeld, N., & Woese, C. (2011). Life is physics: Evolution as a collective phenomenon far from equilibrium. *Annual Review of Condensed Matter Physics, 2*, 375–399.

Gordon, G. J. (2019). Environmental personhood. *Columbia Journal of Environmental Law, 43*(1), 49. https://doi.org/10.7916/cjel.v43i1.3742

Grice, E. A., & Segre, J. A. (2011). The skin microbiome. *Nature Reviews Microbiology, 9*(4), 244–253.

Grilli, M. D., & Verfaellie, M. (2015). Supporting the self-concept with memory: Insight from amnesia. *Social Cognitive and Affective Neuroscience, 10*, 1684–1692.

Guevara, R., Mateos, D., & Perez Velazquez, J. L. (2020). Consciousness as an emergent phenomenon: A tale of different levels of description. *Entropy, 22*, 921. https://doi.org/10.3390/e22090921

Haggard, P. (2008). Human volition: Towards a neuroscience of will. *Nature Reviews. Neuroscience, 9*, 934–946. https://doi.org/10.1038/nrn2497

Hallett, M. (2007). Volitional control of movement: The physiology of free will. *Clinical Neurophysiology, 118*, 1179–1192.

Horowitz, A. (2017). Smelling themselves: Dogs investigate their own odours longer when modified in an "olfactory mirror" test. *Behavioural Processes, 143*, 17–24.

Hume, D. (1740). In P. H. Nidditch & L. A. Selby-Bigge (Eds.), *Treatise of human nature*. Oxford University Press, 1978.

Kauffman, S. A. (1993). *The origins of order: Self-organization and selection in evolution*. Oxford University Press.

King, T. (1996). The one about coyote going west. In S. Kamboureli (Ed.), *Making a difference: Canadian multicultural literature*. Toronto University Press.

Klein, S. B., & Nichols, S. (2012). Memory and the sense of personal identity. *Mind, 121*, 677–702.

Kokarovtseva, L., Jaciw-Zurakiwsky, T., Arbocco, R. M., Frantseva, M. V., & Perez Velazquez, J. L. (2009). Excitability and gap junction-mediated mechanisms in nucleus accumbens regulate self-stimulation reward in rats. *Neuroscience, 159*, 1257–1263.

Kornhuber, H. H., & Deecke, L. (1964). Bereitschaftspotential, EEG activity preceding volitional movements. *Pflügers Archiv, 281*, 52.

Kramar, M., & Alim, K. (2021). Encoding memory in tube diameter hierarchy of living flow network. *Proceedings of the National Academy of Sciences, 118*(10), e2007815118.

Kubota, J. T., Banaji, M. R., & Phelps, E. A. (2012). The neuroscience of race. *Nature Neuroscience, 15*, 940–948. https://doi.org/10.1038/nn.3136

Lambert, F. L. (2002). Disorder –A cracked crutch for supporting entropy discussions. *Journal of Chemical Education, 79*(2), 187.

Lanciego, J. L., Luquin, N., & Obeso, J. A. (2012). Functional neuroanatomy of the basal ganglia. *Cold Spring Harbor Perspectives in Medicine, 2*(12), a009621. https://doi.org/10.1101/cshperspect.a009621

Laurent, G. (2002). Olfactory network dynamics and the coding of multidimensional signals. *Nature Review Neuroscience, 3*, 884–895. https://doi.org/10.1038/nrn964

Lenkei, R., Faragó, T., Zsilák, B., et al. (2021). Dogs (*Canis familiaris*) recognize their own body as a physical obstacle. *Scientific Reports, 11*, 2761. https://doi.org/10.1038/s41598-021-82309-x

Libet, B., Gleason, C. A., Wright, E. W., & Pearl, D. K. (1983). Time of conscious intention to act in relation to onset of cerebral activity (readiness-potential): The unconscious initiation of a freely voluntary act. *Brain, 106*, 623–642. https://doi.org/10.1093/brain/106.3.623

Locke, J. (1690). *An essay concerning humane understanding* (1st ed.). Thomas Basset.

Lopes da Silva, F. (2003). Visual dreams in the congenitally blind? *Trends in Cognitive Sciences, 7*(8), 328–330.

McGinn, C. (1989). Can we solve the mind-body problem? *Mind, 48*, 348–366.

Melzack, R. (2001). Pain and the neuromatrix in the brain. *Journal of Dental Education, 65*(12), 1378–1382.

Merker, B. (2005). The liabilities of mobility: A selection pressure for the transition to consciousness in animal evolution. *Consciouness & Cognition, 14*, 89–114.

Merleau-Ponty, M. (1945). *Phénoménologie de la perception*. Gallimard.

Mesulam, M. M. (1981). Dissociative states with abnormal temporal lobe EEG—Multiple personality and the illusion of possession. *Archives of Neurology, 38*(3), 176–181.

Miller, L. (2000). The predator's brain: Neuropsychodynamics of serial killing. In *Serial offenders: Current thought, recent findings* (pp. 135–166). CRC Press.

Miller, G. (2009). On the origin of the nervous system. *Science, 325*, 24–26. https://doi.org/10.1126/science.325_24

Milner, A. D., Perrett, D. I., Johnston, R. S., et al. (1991). Perception and action in visual form agnosia. *Brain, 114*, 405–428.

Minsky, M. (1985). *The society of mind*. Simon & Schuster.
Monsó, S. (2019). How to tell if animals can understand death. *Erkenntnis*. https://doi.org/10.1007/s10670-019-00187-2
Northoff, G., & Bermpohl, F. (2004). Cortical midline structures and the self. *Trends in Cognitive Sciences, 8*, 102–107.
Parlikar, R., Bose, A., & Venkatasubramanian, G. (2019). Schizophrenia and corollary discharge: A neuroscientific overview and translational implications. *Clinical Psychopharmacology and Neuroscience, 17*(2), 170–182. https://doi.org/10.9758/cpn.2019.17.2.170
Perez Velazquez, J. L. (2009). Finding simplicity in complexity: General principles of biological and nonbiological organization. *Journal of Biological Physics, 35*, 209–221.
Perez Velazquez, J. L. (2020). On the emergence of cognition: From catalytic closure to neuroglial closure. *Journal of Biological Physics, 46*(1), 95–119.
Perez Velazquez, J. L., & Estella Orero, C. B. (2020). *DBS—An Odyssey through the mind*. Amazon.com Services LLC.
Perez Velazquez, J. L., & Frantseva, M. V. (2009). *The brain-behaviour continuum—The subtle transition between sanity and insanity*. World Scientific.
Perez Velazquez, J. L., & Wennberg, R. (2004). Metastability of brain states and the many routes to seizures: Numerous causes, same result. In *Recent research developments in biophysics*, Recent Res. Devel. Biophys (Vol. 3, pp. 25–59). Transworld Research Network.
Perez Velazquez, J. L., Mateos, D. M., & Guevara Erra, R. (2019). On a simple general principle of brain organization. *Frontiers in Neuroscience, 13*, 1106. https://doi.org/10.3389/fnins.2019.01106
Petoft, A. (2015). Neurolaw: A brief introduction. *Iranian Journal of Neurology, 14*(1), 53–58.
Piolino, P., Hannequin, D., Desgrangesm, B., Girard, C., Beaunieux, H., Giffard, B., Lebreton, K., & Eustache, F. (2005). Right ventral frontal hypometabolism and abnormal sense of self in a case of disproportionate retrograde amnesia. *Cognitive Neuropsychology, 22*, 1005–1034.
Poincaré, H. (1917). *La Science et l'hypothèse*. Flammarion.
Prigogine, I. (1980). *From being to becoming: Time and complexity in the physical sciences*. W. H. Freeman & Co Ltd..
Ring, K., & Cooper, S. (1997). Near-death and out-of-body experiences in the blind: A study of apparent eyeless vision. *Journal of Near-Death Studies, 16*(2), 101–147.
Ring, K., & Cooper, S. (2008). *Mindsight: Near-death and out-of-body experiences in the blind*. iUniverse, Inc..
Rogers, D. S., Deshpande, O., & Feldman, M. W. (2011). The spread of inequality. *PLoS One, 6*(9), e24683. https://doi.org/10.1371/journal.pone.0024683
Rolls, E. T. (1999). *The brain and emotion*. Oxford University Press.
Roth, G., & Dicke, U. (2005). Evolution of the brain and intelligence. *Trends in Cognitive Sciences, 9*, 250–257.
Sapolsky, R. M. (2017). *Behave: The biology of humans at our best and worst*. Penguin Press.
Schall, J. D. (2001). Neural basis of deciding, choosing and acting. *Nature Reviews Neuroscience, 2*, 33–42.
Schiff, N., Giacino, J., Kalmar, K., et al. (2007). Behavioural improvements with thalamic stimulation after severe traumatic brain injury. *Nature, 448*, 600–603.
Schlichting, H. J., & Nordmeier, V. (1996). Strukturen im Sand—Kollektives Verhalten und Selbstorganisation bei Granulaten Math. *Naturwissenschaften, 49*, 323–332.
Searle, J. (1992). *The rediscovery of the mind*. MIT Press.
Shettleworth, S. J. (1998). *Cognition, evolution, and behavior*. Oxford University Press.
Shinbrot, T., & Muzzio, F. J. (2001). Noise to order. *Nature, 410*, 251–258.
Simard, S. W., Beiler, K. J., Bingham, M. A., Deslippe, J. R., Philip, L. J., & Teste, F. P. (2012). Mycorrhizal networks: Mechanisms, ecology and modelling. *Fungal Biology Reviews, 26*(1), 39–60. https://doi.org/10.1016/j.fbr.2012.01.001
Slater, M., Pérez Marcos, D., Ehrsson, H., & Sanchez-Vives, M. V. (2009). Inducing illusory ownership of a virtual body. *Frontiers in Neuroscience, 3*(2), 214–220. https://doi.org/10.3389/neuro.01.029.2009

Smith, J. M., & Price, G. R. (1973). The logic of animal conflict. *Nature, 246*, 15–18.

Sornette, D. (2000). *Critical phenomena in natural sciences*. Springer Verlag.

Sperry, R. W. (1952). Neurology and the mind-brain problem. *American Scientist, 40*, 291–312.

Swain, J. (2002). Quantum mechanics, computability theory and life. In H. P. Durr, F. A. Popp, & W. Schommer (Eds.), *What is life? Scientific approaches and philosophical positions* (pp. 329–342). World Scientific.

Tricarico, E., Borrelli, L., Gherardi, F., & Fiorito, G. (2011). I know my neighbour: Individual recognition in Octopus vulgaris. *PLoS One, 6*(4), e18710. https://doi.org/10.1371/journal.pone.0018710

Vallverdú, J., Castro, O., Mayne, R., Talanov, M., Levin, M., Baluška, F., Gunji, Y., Dussutour, A., Zenil, H., & Adamatzky, A. (2018). Slime Mould: The fundamental mechanisms of biological cognition. *Biosystems, 165*, 57–70. https://doi.org/10.1016/j.biosystems.2017.12.011

Varela, F. J. (1996). Neurophenomenology: A methodological remedy for the hard problem. *Journal of Consciousness Studies, 3*(4), 330–349.

Varela, F. J., Thompson, E., & Rosch, E. (1991). *The embodied mind*. MIT Press.

von Hayek, F. (1952). *The sensory order*. University of Chicago Press.

Voss, U., Holzmann, R., Tuin, I., & Hobson, A. J. (2009). Lucid dreaming: A state of consciousness with features of both waking and non-lucid dreaming. *Sleep, 32*, 1191–1200. https://doi.org/10.1093/sleep/32.9.1191

Walsh, J. M., Doyle, L. W., Anderson, P. J., Lee, K. J., & Cheong, J. L. Y. (2014). Moderate and late preterm birth: Effect on brain size and maturation at term-equivalent age. *Radiology, 273*(1), 232–240.

Wennberg, R., McAndrews, M.-P., Zumsteng, D., & Perez Velazquez, J. L. (2009). The sign of the cross as a learned automatism? *Epilepsy & Behavior, 15*(3), 394–398.

Wheeler, J. A. (1977). Genesis and observership. In R. E. Butts & J. Hintikka (Eds.), *Foundational problems in the special sciences*. D. Reidel Publishing Company.

Zhu, Z., et al. (2012). Corrigendum: A multi-omic map of the lipid-producing yeast Rhodosporidium toruloides. *Nature Communications, 3*, 1112.

Index

J. L. Perez Velazquez, V. Nenadovic, *Being and becoming*,
https://doi.org/10.1007/978-3-030-78264-1

GPSR Compliance

The European Union's (EU) General Product Safety Regulation (GPSR) is a set of rules that requires consumer products to be safe and our obligations to ensure this.

If you have any concerns about our products, you can contact us on ProductSafety@springernature.com

In case Publisher is established outside the EU, the EU authorized representative is:

Springer Nature Customer Service Center GmbH
Europaplatz 3
69115 Heidelberg, Germany

Batch number: 10722209

Printed by Printforce, the Netherlands